THE PHYSICS OF FORAGING
An Introduction to Random Searches and Biological Encounters

Do the movements of animals, including humans, follow patterns that can be described quantitatively by simple laws of motion? If so, then why? These questions have attracted the attention of scientists in many disciplines and stimulated debates on ecological matters to queries such as, "how can there be free will if one follows a law of motion?"

This is the first book on this rapidly evolving subject, which introduces random searches and foraging in a way that can be understood by readers with no previous background on the subject. It reviews theory as well as experiments, addresses open problems and perspectives, and discusses applications ranging from the colonization of Madagascar by Austronesians to the diffusion of genetically modified crops.

The book will interest physicists working in the field of anomalous diffusion and movement ecology as well as ecologists already familiar with the concepts and methods of statistical physics.

GANDHIMOHAN M. VISWANATHAN is Professor of Physics at the Universidade Federal do Rio Grande do Norte. Previously, he was Associate Professor at the Instituto de Física, Universidade Federal de Alagoas, where he spent more than a decade investigating the complex phenomenology seen in physical, biological, and economic systems.

MARCOS G. E. DA LUZ is Associate Professor in the Departamento de Física, Universidade Federal do Paraná. He works with numerical and analytical methods in disorder, complexity and nonlinearity in classical and quantum systems, and with theoretical models in ecology.

ERNESTO P. RAPOSO is Associate Professor in the Laboratório de Física Teórica e Computacional, Departamento de Física, Universidade Federal de Pernambuco. His current research interests include statistical physics of random searches and the foraging problem, disordered antiferromagnets, and quantum field theory of quasi-unidimensional magnetic polymer chains.

H. EUGENE STANLEY is University Professor and Director of the Center for Polymer Studies, Boston University; Professor of Physics, Chemistry, and Biomedical Engineering; and Professor of Physiology at Boston University School of Medicine. A pioneer in interdisciplinary science, he has won the Boltzmann Medal and numerous other awards for his contributions to this field.

THE PHYSICS OF FORAGING

An Introduction to Random Searches and Biological Encounters

GANDHIMOHAN M. VISWANATHAN

Universidade Federal do Rio Grande do Norte, Brazil

MARCOS G. E. DA LUZ

Universidade Federal do Paraná, Brazil

ERNESTO P. RAPOSO

Universidade Federal de Pernambuco, Brazil

H. EUGENE STANLEY

Boston University, USA

CAMBRIDGE
UNIVERSITY PRESS

CAMBRIDGE UNIVERSITY PRESS
Cambridge, New York, Melbourne, Madrid, Cape Town,
Singapore, São Paulo, Delhi, Tokyo, Mexico City

Cambridge University Press
The Edinburgh Building, Cambridge CB2 8RU, UK

Published in the United States of America by Cambridge University Press, New York

www.cambridge.org
Information on this title: www.cambridge.org/9781107006799

First published 2011

Printed in the United Kingdom at the University Press, Cambridge

A catalog record for this publication is available from the British Library.

Library of Congress Cataloging in Publication data
The physics of foraging : an introduction to random searches and
biological encounters / Gandhimohan M. Viswanathan . . . [et al.].
p.   cm.
Includes bibliographical references and index.
ISBN 978-1-107-00679-9 (hardback)
1. Animal behavior – Mathematical models.   2. Home range (Animal geography) –
Mathematical models.   3. Biological invasions – Mathematical models.   4. Animal ecology –
Mathematical models.   I. Viswanathan, Gandhimohan M.   II. Title.
QL751.65.M3P49   2011
591.5 – dc22   2011002166

ISBN 978-1-107-00679-9 Hardback

Dedicated to our families

Contents

Preface

As the FBI helps a 14-year-old victim who escaped from a dangerous polygamist self-proclaimed prophet, it is faced with the question of how to search 2200 square miles of mountain desert.

"How rough is the terrain? Because the rougher the terrain, the more likely she was forced into a Lévy flight type movement. I can create a viable search pattern," says Charlie Eppes, the mathematical genius.

"It's like when you lose your keys," explains Amita Ramanujan, his girlfriend and former doctoral student. "You don't methodically search every inch of your house from front to back. You look like crazy in one area, and then jump to the next most likely area and look there."

The preceding dialogue, from the American television series *Numb3rs*, shows how far the theory of Lévy flight foraging has penetrated mainstream science. Although the term *foraging* has a biological connotation, in fact, biological foraging is a special case of random searches. Michael Shlesinger, for instance, has pointed out the relevance of random searches to operations research in World War II, involving the hunt for enemy submarines.

There are intriguing aspects of the random search problem that are peculiar to biological foraging. Why should the movements of freely moving animals follow any natural law at all? This is a fascinating question, and we find it remarkable that animals – and even humans – that possess a degree of "free will" actually move in a manner that can be described quantitatively by physical principles.

Questions such as how they move (and why) have attracted the attention of physicists for a number of other reasons as well. The trajectories of individual organisms as they forage and search closely resemble certain kinds of random walks studied by theoretical physicists. Encounter interactions of moving organisms thus have a close parallel with reaction-diffusion processes seen in nonequilibrium statistical mechanics. These mathematical similarities and the abundance of experimental data have allowed physicists to make contributions toward the study of foraging. Hence the choice of title, *The Physics of Foraging.*

This book is an introduction to the interdisciplinary study of how organisms move in order to encounter and interact. Our goal is to bring together relevant theoretical concepts, along with a review of the experimental findings, to allow the current literature to be readable and understandable. The focus is on the statistical properties of the trajectories of single organisms. Such quantities are statistically coercive, making them useful in constraining the range of possible behaviors or in predicting the conditions under which the adoption of a given random walk strategy might confer some relative advantage, and so forth.

The statistical physics approach to the problem of foraging and random searches in many ways complements the more traditional treatment given in other disciplines. Collaboration between biologists and ecologists, on one hand, and physicists, on the other, began in earnest about two decades ago. By the early 1990s, there was enough empirical evidence to suggest that organisms do not always diffuse like particles performing Brownian motion. In collaboration with Sergey Buldyrev and Shlomo Havlin, we proposed in 1999 what has become known as the Lévy flight foraging hypothesis to try to account for the observed phenomena. Since then, a number of works coauthored by physicists have further impacted the interdisciplinary subfield of theoretical movement ecology. At present, the field is still evolving rapidly; hence we also discuss in the book a number of open problems and perspectives, both from theoretical and empirical points of view.

We are in great debt to many people with whom, along the years, we have discussed many aspects of random searches, Lévy flights, and related topics. We thank our key collaborators in the two first *Nature* papers that congealed our interest in the topic: V. Afanasyev, S. V. Buldyrev, S. Havlin, E. J. Murphy, and the late P. A. Prince. We thank Ivars Peterson for the appealing article on the first *Nature* paper that helped focus attention on statistical physics approaches to understanding the more enigmatic aspects of foraging. We also thank Mark Buchanan for drawing considerable attention to the topic in a more recent feature article in *Nature*. We have had discussions with many other people to whom we are grateful: J. S. Agnaldo, F. Bartumeus, M. W. Beims, D. Boyer, C. Carvalho, J. Catalan, S. Cavalcante, M. D. Coutinho-Filho, J. C. Cressoni, A. Davis, A. M. Edwards, C. L. Faustino, M. L. Felisberto, N. M. Freeman, U. L. Fulco, M. Gitterman, L. Giuggioli, I. M. Gléria, H. D. Jennings, V. M. Kenkre, J. Klafter, A. Y. Kasakov, L. S. Lucena, M. L. Lyra, A. Marshak, J. L. Mateos, R. Metzler, O. Miramontes, C. M. Nascimento, A. M. Nemirovsky, R. W. Nowak, E. J. Nunes-Pereira, F. S. Passos, R. A. Phillips, M. C. Santos, M. F. Shlesinger, L. R. da Silva, M. A. A. da Silva, I. M. Sokolov, C. Tsallis, T. M. Viswanathan, and N. W. Watkins. Moreover, we are in great debt for the support and encouragement provided by our families. Last, but by no means least, we thank Dr. Simon Capelin of Cambridge University Press, and the entire CUP team, for managing all aspects involved in the editing and production of this book.

The research that has culminated in this book would have been impossible to undertake without financial support. We acknowledge BNB, CAPES, CNPq, CT-Energ/FINEP, CT-Infra/FINEP, FACEPE, FAPEAL, Fundação Araucária, and NSF for funding.

Part I

Introduction: Movement

1

Empirical motivation for studying movement

1.1 How do organisms really move, and why?

Animals must move in order to eat and mate. They may also need to escape their predators. The details of their movement may depend on many factors such as climate, temperature, concentrations of pheromones, or the local density of other organisms (including humans) [172, 365]. Although such factors may affect the velocity, sinuosity, or specific trajectory taken, they do not change the primary reasons underlying the movement: the biological necessity of interactions or "encounters" with other organisms.

Given the ubiquity of moving organisms, a number of important questions arise naturally. For example, the priority in the order of driving factors that determine animal movement is not yet well understood and may even depend on the specific activity an organism is performing at a given time. However, there has been progress in understanding *how* organisms move, i.e., what patterns the trajectories follow. This is the main focus of this book. As an illustrative example, Figure 1.1 shows how spider monkeys move in the Yucatan Peninsula when they are allowed to roam freely. What factors determine the shape and the statistical properties of such trajectories? If we know the answers to these questions, we can venture beyond phenomenological descriptions and ask about causation: for a specific species of organism, why do the organisms actually move as they do? *Cui bono*,[1] i.e., what advantage or benefit do they gain from such behavior? Moreover, how did the specific biological mechanisms involved in generating the behavior evolve?

Such questions have prompted investigations in the new interdisciplinary subfield known as *movement ecology*. Because these questions are relevant in such research areas as random walk theory, stochastic processes, and anomalous

---

[1] The Latin adage, suggestive of hidden motives, asks, "for whose benefit?"

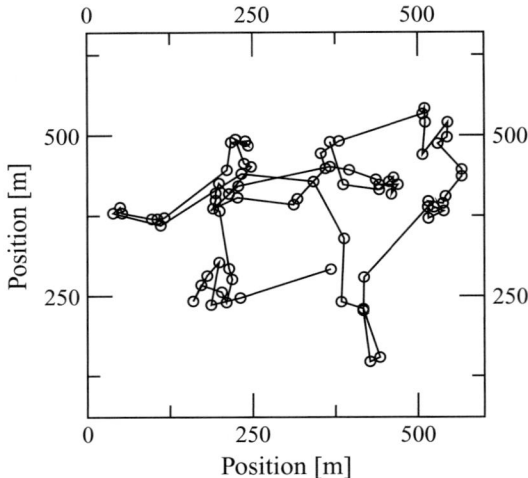

Figure 1.1 Movement patterns of spider monkeys in the Yucatan Peninsula, studied and reported by Ramos-Fernández *et al.* [292]. Patterns generated by moving animals share features with random walks studied by physicists. Compare the trajectory with those of humans, shown in Figure 7.1. The qualitative and quantitative similarities seen in the patterns of diverse animals might be because for all or most animals, the temporal and spatial organization of movement patterns serves to optimize physical quantities such as encounter rates and search efficiencies.

diffusion, they have also attracted the attention of physicists (see [18, 45, 49, 136, 159, 170, 232, 238, 247, 249, 254, 269, 274, 295, 303, 324, 340] appearing in a special issue [91] of *Journal of Physics A*, published in October 2009 and dedicated to the random search problem).

1.2 Biological encounters as a reaction-diffusion process

Biological encounters typically involve a diffusive (i.e., transport) component and a reactive (i.e., interaction) component, such as eating or mating. They thus represent a special case of reaction-diffusion processes. Typically, the diffusion process is linear in the sense that the superposition principle holds for the probability density functions of the random walkers.

In this case, the superposition principle guarantees that the probability of finding one of many random walkers at a specified position will equal the sum of the probabilities of finding each of them individually at that position. In more technical terms, the superposition principle guarantees the existence of random walk propagators.

However, for the superposition principle to hold, the random walkers must not interact among each other because such interactions will typically lead to nonlinear effects. (Noninteracting random walkers must always obey linear Fokker-Planck equations for the probability density function for the walkers [60, 193].) This linear approach to diffusion is remarkably useful. For example, Sparrevohn *et al.* [356] have found that thousands of fish released at a single point diffuse as random walkers once the movement of the water (i.e., advection) is taken into account.

In contrast, the reaction process necessarily involves one "particle" interacting with another, which can lead to nonlinear phenomena. Consider, for example, "reactions" represented by a predator wanting to eat its prey. Whereas two meals of prey may, in principle, be approximately twice as beneficial as a single meal, 10^3 meals are not approximately 10^3 times more beneficial. Reactions between predator and prey thus inherently deviate from linear behavior.

Eating, mating, and pollination represent distinct reactions. To a large extent, such biological interactions fall into two general categories. The first category includes interspecific interactions, typically, a trophic interaction between a consumer and a consumable, which can adopt the form of predation, parasite infection, or mutual rewarding (e.g., flowers and pollinators). The second category relates to interactions between individuals of the same species, e.g., mating or territorial competition. Two-species reaction-diffusion models (i.e., those with two reacting species) can thus be used to describe many ecological systems [19].

Most importantly, the diffusion (i.e., movement) that accompanies such diverse reactions remains the same, at least in a first approximation. Specifically, we do not expect the randomness seen in the movements to depend strongly on whether the organism is foraging for food or searching for a mate (or for something else), so long as such relevant search cues as the density of organisms remain comparable. This premise, to the extent that it remains approximately valid, justifies the study of the diffusive properties of biological encounter processes independently of the nature of the reactive processes. In this book, we focus mainly on the encounter rates between organisms; i.e., we address only the diffusive aspects of the underlying reaction-diffusion process.

This approach can be tailored to take into account new types of behaviors. For example, search for food may not necessarily be dominant. Avoidance of predators may also be important [115]. A predating organism may benefit from increased encounter rates with its prey, while simultaneously benefiting from lower encounter rates with its own predators.

Factors conditioning encounter rates between organisms are believed to play a crucial role in the ecological constraints important in the evolution of life. These interactions can involve many potential factors and multiple ecological adaptive pathways. In this context, it is difficult to exaggerate the importance of movement. For example, locomotion and its detection go hand in hand, and therefore it has been hypothesized that the evolution of external bilateral sensory organs (e.g., eyes and ears) is partially due to the sudden increase in spatial complexity and patchiness of the marine odor landscape during the Ediacaran-Cambrian interval [289], about 542 million years ago. Foraging and search strategies represent one of the most important factors affecting encounter rates. We thus wonder whether they might even have influenced indirectly the evolution of the sensory apparatus.

We study encounter rates in a framework [19] that distinguishes between two kinds of interacting organisms. The organism is either a *searcher*, e.g., forager, predator, parasite, pollinator, or the active gender in the search activity involved in the mating process, or it is a *target*, e.g., prey, food, or the passive gender in the mating activity. Statistical models of foraging do not need to take into consideration the "microscopic" details of the process – they are essentially irrelevant to the averages. In this sense, it is important to recognize the limitations and applicability of such models. Despite this "coarse graining," these models lead to statistically robust results, precisely because they do not depend on the particular biological implementation of the search mechanisms. There is a long tradition in statistical physics in which apparently simple models lead to remarkably good agreement with experiment (e.g., the Ising model of ferromagnetic phase transitions). We will return to this topic in Section 2.5, and again in Part III.

The framework we adopt allows for considerable variation and easily generalizes to cover new cases. For example, the search could be guided almost entirely by external cues, either by the cognitive (memory) or detection (olfaction, vision, etc.) skills of the searcher, or the searches might not be oriented, thus effectively becoming stochastic processes. Even if the actual process is completely deterministic, a statistical approach is useful, or perhaps even necessary, if the environment is a disordered medium. Deterministic walks (e.g., the traveling salesman problem and the traveling tourist problem [358]) in random environments can appear indistinguishable from (genuinely stochastic) random walks, an issue we discuss in greater detail in Part IV.

One of the important questions in the random search problem can be stated as follows: what is the most efficient strategy for searching randomly located objects whose exact locations are not known *a priori*? The question is relevant because performing efficient searches is not trivial or straightforward. Typically, the searchers have a certain degree of "free will" as they move. (By free will, we refer not to contracausal behavior but rather to the degree of autonomy commonly seen in human beings and other organisms; see Section 14.3.) Nevertheless, organisms are at the same time subject to physical and biological constraints that restrict their modes of action. For example, a predator cannot search too long without finding food or it will perish. There is therefore an interplay between free will and constraints, with somewhat unpredictable consequences, including the possible emergence of complex behavior. Indeed, there is an inherent complexity and dynamical richness

(a) Short-range dispersal

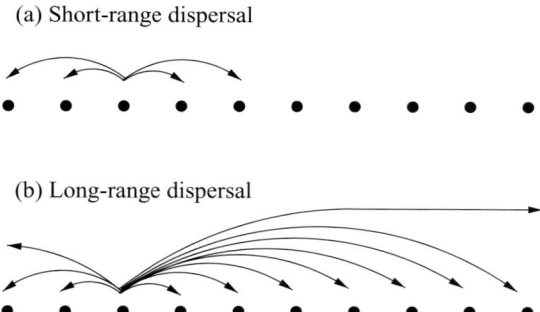

(b) Long-range dispersal

Figure 1.2 (a) The usual uncorrelated Brownian random walk assumes that the random walker can move only to neighboring sites within some limited range. In the long time limit, the mean squared displacement of such random walks grows linearly with time. (b) The behavior is qualitatively different for a random walker that can move to sites arbitrarily far away. In Lévy flights and walks, the probability of moving to a site ℓ units away decays algebraically, i.e., asymptotically, as a negative power of ℓ. The mean squared displacement for such *anomalous diffusion* (see Chapter 3) can grow superlinearly.

in the random search problem – and, by proxy, in other applications involving biological encounters.

1.3 Impact and scientific importance

Mathematical modeling of the movement and diffusion [265] of plant seeds and pollen, animals [378], micro-organisms [35], cells [230], organelles [160], and even genetic material [158] has played an important role in biology, ecology, and medicine [84]. Ignorance of how organisms actually move has motivated scientific research to a large extent, but there are a number of other reasons for the growing interest in the problem. Indeed, in a world in which we have an increasing aware-ness of how organisms, ecosystems, and society are intrinsically interconnected, the question of how and why organisms move becomes important even for political, economic, environmental, and health-related reasons.

Consider, for example, the seemingly "pedantic" question of how pollen dif-fuses [158]. The spread of organisms represents a particular case of the dispersal of genetic material. In this context, Shaw *et al.* [338] report that rapeseed pollen disperses in a manner inconsistent with exponential (e.g., Gaussian) models but con-sistent with the power laws associated with long-range dispersal. Figure 1.2 illus-trates the difference between short- and long-range interactions. The scaling expo-nents for rapeseed pollen lie in the Lévy walk range [338]. Lévy walks represent a special case of superdiffusion, and a detailed description of Lévy processes appears in Chapter 3. Genetically modified (GM) crops of rapeseed resistant to herbicides have been engineered by a multinational biotechnology corporation.[2] In principle, the shape and weight of pollen in GM crops do not differ on any fundamental level from those in non-GM crops. For this reason, superdiffusion of genes from GM crops is possible [338], with potential policy implications. GM crops represent just one example of how the diffusive properties of biological systems can have far-reaching effects. Lévy flights may also find application in efforts to control the spread and proliferation of diseases via eradication strategies for optimal disease containment [103], e.g., vaccination [311]. Indeed, any mechanism that slows the diffusion of a disease agent or vector can be exploited; e.g., the time spent by mosquitoes in search of oviposition targets significantly reduces the reproductive rate of the malaria parasite [142].

We cite a second example, one related to environmental issues and conservation efforts. The complexity of foraging and movement patterns and the rich dynamics underlying the relationships between predator and prey render plausible, if not inevitable, a cascade of environmental impacts due to human activities. It is well known that human fishing activities negatively affect marine ecosystems [24, 276, 382]. Waugh *et al.* [400] have documented the change in behavior of royal albatrosses around fishing operations. Similarly, female leatherback turtles (*Dermochelys coriacea*) in the Atlantic appear to fine-tune their foraging behavior and daily activity patterns to take into account the local conditions [150]. In the context of such foraging flexibility, Hays *et al.* [150] have highlighted the importance of regulating whole-ocean fishing gear to minimize turtle bycatch. A quantitative description of animal movement may well allow for a more nuanced and scientifically informed approach to environmental policy. In a similar context, the diffusive properties of sheep are valuable in identifying and predicting spots of potential land degradation and for planning the distribution of flocks in the context of sustainable management in shrubby rangelands [38].

Indeed, beyond the purely theoretical interest in the movement of animals and organisms in general, awareness has grown about the potential ecological impact arising from the poorly understood interactions between anthropogenic changes in the environment, on one hand, and dispersal, on the other. We cite a third and final example of the importance of studying animal movement: Van Houtan *et al.* [383] have studied the sensitivity to landscape fragmentation of dispersal mechanisms for species of birds in the Amazon tropical forests, in the context of an intriguing problem of great concern – why species disappear from forest fragments. Their results seem to suggest a nonlinear and highly complex set of interactions, with considerable interspecific variability. Birds that disappear from fragments show a

---

[2] Monsanto (NYSE: MON) has manufactured the herbicide glyphosate and has produced GM seeds that grow into plants resistant to this herbicide.

tendency to disperse over large distances in connected forest but not in fragmented forest. In contrast, species that persist in fragments do not cross gaps as often, yet disperse farther after fragmentation than before [383]. The authors conclude that *heavy-tailed* models better explain the dispersal kernels than Gaussian models and that tropical forest birds disperse more than commonly thought.

Other applications transcend biology and ecology. Biological foraging is actually a special case of random search. It can also include, for instance, searches for misplaced keys, missing children, and international criminals. Similarly, enzymes sometimes perform random searches for specific DNA sequences. In the context of operations research, Shlesinger [340] has noted the relevance of random searches to the hunt for submarines in World War II. Indeed, the physics and mathematics underlying random searches are so general, and their applications so diverse, that they became a topic of an American TV drama.[3]

1.4 Follow the data

Significant progress has been made over the last hundred years in the study of linear and nonlinear partial differential equations, nonlinear maps, and chaos [333]. These advances led to theoretical expectations that, until the 1980s, tended to bear on the empirical investigations. But the inherent stochasticity observed in movement data makes it difficult to link movement complexity with dynamical models of population processes [275], for example. Ecological theory traditionally held premises similar to those seen in the theory of equilibrium statistical mechanics for memory-free, scale-specific processes, normal diffusion, and Fickian transport [131]. Gautestad and Mysterud [131] have noted that animals from many taxa generally express strategic homing, site fidelity (i.e., memory), and same-species or conspecific attraction (with possibly non-Fickian transport due to interaction among diffusing individuals). Hence ecological systems and processes have more in common with complex systems far from equilibrium than was suspected earlier. It is no surprise that studies of nonequilibrium statistical mechanics [179, 198], scale-free Lévy processes [209, 342], fat-tailed or leptokurtic [417] dispersal kernels, and non-Markovian chains with long-range memory have contributed to the advance of theoretical ecology (Figure 1.3).

Because the basic premises of equilibrium statistical physics do not necessarily hold for biological and ecological systems, a data-centric scientific research program allows a quantitatively more correct and theoretically less biased approach. Such a program has two guiding principles that represent a break from the earlier

[3] Lévy flight searches are discussed in the episode "Nine Wives" (episode 12 of season 3) of the TV series *Numb3rs*.

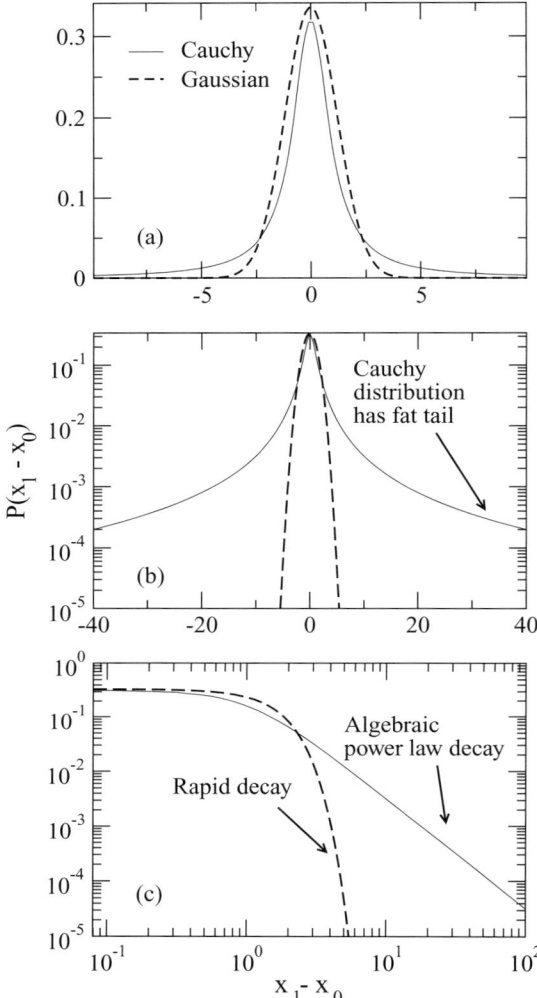

Figure 1.3   (a) Normalized probability density functions to move from position
x_0 at time t_0 to position x_1 at time t_1 for a Brownian random walk and a Lévy
flight, shown for some large but arbitrary value of $t_1 - t_0$. A Cauchy distribution of
random walk jump lengths can be used to generate a Lévy flight. The same curves
are plotted on (b) semilog and (c) log-log scales. These distributions, known as
random walk propagators (see Chapter 3), are Gaussian for Brownian random
walks but have asymptotic power law decay for Lévy flights. For this reason, such
distributions are often described as "fat tailed" and "heavy tailed" in the literature.
They also have scale-free properties (see Figure 2.1).

approach: (1) to discard preconceived expectations relating to the central limit theorem, Gaussian statistics, and Brownian motion and, instead, to expect a rich variety of behaviors; and (2) to construct a theoretical framework that, while testable (i.e., falsifiable) and amenable to model comparison, nevertheless captures the most important statistical properties of the data.

We will argue in subsequent chapters that the formalism of Wiener processes, Gaussian statistics, and the (non-fractional) Laplacian operator in the diffusion (or heat) equation does not adequately describe the experimental findings in a large number of studied cases. The discrepancy between model and data to some extent mirrors the now well-documented inadequacy of Gaussian statistics and Wiener processes to correctly model price fluctuations in financial markets [222] (notwithstanding the Black-Scholes theory of options pricing).

Historically, a major theoretical advance in the study of stochastic processes that could rival Gaussian-like statistics for describing animal movement came from the study of Lévy processes. Paul Lévy generalized the central limit theorem to cover distributions with diverging second moments [209]. The main difference between Wiener and Lévy processes stems from the fact that the probability density function of a Lévy-distributed random variable has infinite variance [209], in contrast to the usual case, for which the variance and all higher moments remain always finite. In the context of animal movement, the earliest reference to the superdiffusion of organisms is a 1986 paper by Shlesinger and Klafter [341] about Lévy random walks.

Today, advances in the experimental methodologies and the abundance of data have greatly increased, qualitatively and quantitatively, our knowledge of how animals move. In addition to very high frequency (VHS) and satellite-tracking telemetry, GPS [278], acoustic telemetry, and geolocation [319], a method known as *dead-reckoning* can also track the underwater fine-scale movement of animals [411]. Dead-reckoning involves using the position of "fix" at an earlier time and advancing it to correct for nonzero velocity.

The data confirm that organisms move with a directional persistence that is absent in Brownian motion. However, they rarely move in perfectly straight lines; hence, Lévy walks cannot perfectly describe how real organisms move. What is most important is that, over the course of two decades of study, a growing consensus among researchers in the relevant fields holds the view that many organisms diffuse anomalously: real organisms rarely have Gaussian probability functions for displacements with a variance that grows linearly in time. Rather, the motion of organisms is rich in variety and surprisingly complex. Presently, it appears that no single *universal* natural law applies across all species: animals with home ranges [135, 293], for example, behave differently from those that migrate seasonally.

The data suggest questions that lie beyond the scope of this book. For example, we will not examine the coherent motion observed in flocks or herds of animals and other "self-propelled particles" [2, 280, 386] (for collective searches, see [232, 324]). We also will not relate the movement of individual organisms to populations or to treat the individual and population levels together [257]. For example, Petrovskii *et al.* [283] argue that ignoring inherent population structure leads to the prediction of Gaussian, or "thin," tails in dispersal curves and that the empirically observed "fat tails" reflect such structure.

In this volume, we limit our discussion to the kinematic description of the trajectory of individual organisms and to the causes that underlie this motion because such a description is already a complex issue. For instance, the locomotion of animals does not always involve aspects only associated with search or encounter processes. Directed walks, for instance, may be related to navigation with respect to external references or internal directional sensory information of body rotations [82, 83] and may not simply be a by-product of a random search.

1.5 Beyond model comparison

An important aspect of a data-centric approach to animal movement involves fitting theoretical curves to empirical data and the consequent comparison of different models. Typically, one asks questions such as which of two given models better describes a particular set of data. Much more importantly, however, we recall that models make assumptions and that these assumptions can lead to (sometimes new) theoretical predictions. Such considerations become especially important given the fact that human perception is based on limited scales of experience, such that often it is difficult to quantitatively describe dispersal and other ecological processes occurring over long time periods and large areas [154].

Beyond model comparison, an empirically motivated approach to the problem requires us to revisit the theoretical underpinnings that we often take for granted. For example, several orders of magnitude separate the theoretical predictions of the time necessary for our human ancestors to have dispersed (i.e., diffused) out of Africa to populate remote and inaccessible parts of the planet. The idea that human beings dispersed like particles performing Brownian motion or the correlated walks associated with small-angle scattering does not agree with known facts (see Chapter 7).

Nevertheless, even this program of "following the data," which lies at the heart of scientific methodology, should be undertaken with some care. When trying to understand a complex process in nature, the empirical evidence is essential. Hypotheses must be compared with the actual data, but the empirical evidence itself may have limitations; that is, it may not be sufficiently large or accurate either to confirm or rule out hypotheses, models, explanations, or assumptions, even when the most sophisticated analytical tools available are used. For movement patterns, this issue may be in some cases particularly delicate. A data set describing the locomotion of an animal is, in some sense, a summary of all the activities in which the animal has been involved during data collection and these activities must be taken into account when analyzing and modeling actual field behavior.

In the following chapters, we make the case that organisms generally diffuse anomalously. We hope that the growing empirical evidence, together with theoretical considerations, will convince the reader of the need to reconsider the traditional view of how organisms diffuse.

2

Statistical physics of biological motion

2.1 Optimal foraging theory

Traits that allow individuals to forage more efficiently can be expected to be naturally selected. The hypothesis that natural mechanisms should drive foraging organisms to maximize their energy intake gave rise to what became known as *optimal foraging theory*. The idea can be traced to studies undertaken by MacArthur and Pianka [219] and Emlen [109] in 1966.

Optimal foraging theory predicts that foragers will behave to maximize the net caloric gain per unit time of foraging. It assumes differentiated functional classes of predators (grazers, parasites, etc.) and provides insight into correlations between physiological features and predation skills (e.g., digestion and ingestion rates). It also highlights the importance of handling time (e.g., for killing and eating prey) [156, 190, 191, 192, 218, 267].

A large body of theoretical work [162, 171] grew in an attempt to deal with the multitude of determinant factors and in order to identify the relevant parameters involved in the predicted optimization [328]. An important example is the marginal value theorem [75, 76], which states that for the forager to maximize the net energy gain per unit time while foraging in a (more or less uniformly) patchy environment, the forager must leave a given patch when the expected net gain from staying in the patch drops to the expected net gain from traveling to (and starting to search in) the next patch. Other important components of classic optimal foraging theory include central place foraging theory [156, 267], competition models (e.g., interference foraging and buffer effects [89, 138, 139]), foraging in population ecology [367], and, more recently, stochastic dynamic programming tools [157].

Optimal foraging theory, as it was conceived within behavioral ecology and presented within Krebs and Davies's seminal book [190, 191], ran into some difficulties in the 1980s. Behavioral ecology as a subject had, to some extent, become overrun by work on sexual selection. The early progress of classic optimal foraging theory to explain and predict animal foraging was beginning to slow. We now know that, although the theoretical framework of optimal foraging theory was well developed intellectually, as a practical framework it did not lend itself to realistic testing. For example, it became necessary to parameterize many aspects of foraging. Similarly, one needed to guess the so-called currencies that animals might minimize or maximize (e.g., net energy gain per unit time or targets found per unit time). Moreover, the test environments were focused on temporal aspects rather than on a fully spatiotemporal description. Finally, classic optimal foraging theory did not take into account the role of missing information. What strategy must a forager choose without full access to all the relevant information (e.g., marginal values for energy gain)? This question immediately suggests a statistical or probabilistic approach.

Perhaps even more significantly, when experimental tests showed that animals do not always optimize the foraging behavior; interest in optimal foraging theory dropped, but new ideas emerged suggesting trade-offs between optimal foraging and predator risk, for example. Biology is replete with suboptimal implementations. For a very interesting review containing different perspectives on optimal foraging theory, see Stephens *et al.* [365].

Since the 1990s, optimal foraging theory has come under the influence of ideas and methods found in statistical physics [18]. The advances made possible by this new interdisciplinary approach have reduced the number of relevant (model) parameters and allowed easier experimental testing – both in the lab and in the field. Bartumeus and Catalan [18] have recently incorporated the general encounter problem into foraging theory. By sketching the assumptions of optimal foraging theory and summarizing recent results on random search strategies, they have suggested ways to extend classic optimal foraging theory into a more general framework.

A number of ideas from statistical physics can be used to study foraging, e.g., disorder, scaling, and universality. In the following sections, we explore how statistical physics can complement and extend classic optimal foraging theory.

2.2 Microscopic versus macroscopic levels of description

How can statistical physics help quantitatively describe complex systems such as those found in ecosystems? The goal of Boltzmann's original research program was to reconcile the laws of thermodynamics with those of classical (i.e., Newtonian) mechanics. Statistical physics focuses on macroscopic phenomena that result from microscopic interactions among innumerable individual components. Consider the motion of an inanimate object such as a cork tossed into a turbulent fluid. The macroscopic behavior of a cork on the ocean depends on the sum total of all collisions that it suffers each picosecond from about 10^{10} microscopic water molecules, but the study of fluid dynamics and turbulence does not depend on summing up these 10^{10} collisions on a computer. Instead, turbulent fluids are understood in terms of a collective phenomenon that does not appeal directly to Newton's laws but rather to the scaling laws of turbulence. An analogous problem arises in many investigations undertaken in biology and ecology [141]. Concepts and techniques that originated in statistical physics have become useful tools for the quantitative analysis of complex systems [361].

In this context, a key question about foraging concerns how the macroscopic statistical properties of the search trajectories relate to the microscopic steps taken by the animal or organism. Global properties such as search efficiencies depend on the macroscopic quantities, whereas at any given time, the organism has access only to the information in its own limited (microscopic) vicinity. Therefore, statistical physics is well suited to the study of foraging as well as other complex phenomena studied in ecology and biology [18].

2.3 Disorder and incomplete information

A second reason for the relevance of statistical physics in the context of biological foraging and random searches concerns entropy (or information), disorder, and stochasticity. A statistical approach [275] is the only option if one does not have access to the microscopic data (e.g., the initial conditions). Indeed, one cause underlying the richness of the foraging problem concerns the "ignorance" of the locations of the randomly located *targets*. Contrary to conventional wisdom, however, the lack of complete information does not necessarily lead to greater complexity. The case of complete information is an illustrative example. If the positions of all target sites are known in advance, then the question of the sequential order in which to visit the sites to reduce the energy costs of locomotion becomes challenging: the famous traveling salesman optimization problem [358, 395]. Not knowing the target site locations, however, considerably modifies the problem and renders it unsolvable by deterministic computational methods [90]. This ignorance-induced stochasticity makes *guessing* unavoidable. The random search problem therefore requires search algorithms that use some element of randomness. Thus a statistical approach to the search problem can take into account the element of ignorance. In other words, incomplete information leaves the search problem underdetermined, such that probabilistic or stochastic strategies become advantageous (if not unavoidable).

It is clear that an organism can choose between two or more equally good solutions only stochastically. Even human decision making can exploit probabilistic strategies [287]. How do you choose between two equally good dinner menu options in a restaurant? Flip a coin, or ask a waiter for his *a priori* unknown recommendation. This point merits careful attention because it goes to the heart of the issue of randomness in animal movement. *The Jungles of Randomness: A Mathematical Safari* [282] discusses randomness and includes descriptions of some counterintuitive and surprising phenomena.

2.4 Scaling and universality

Two concepts central to statistical physics are scaling and universality. These two conceptual "pillars" tell us how complex systems formed of interacting subunits should behave [357]. These conceptual pillars were constructed more than a quarter century ago by scientists studying the behavior of a system near its critical point. Progress on this was made possible by a combination of experiment and phenomenological theory.

To help conceptualize the problem and establish the main ideas, the Ising model came to play a key role. Indeed, most of the ideas that emerged were tested on this model system, whose behavior is well understood, thanks, in part, to Onsager's brilliant exact solution of the model in two dimensions [266]. This achievement perhaps began one of the most exciting chapters in the history of statistical physics, and we refer the interested reader to the standard texts on the subject.

The Ising model is defined to be a set of classical spins localized on the sites of a lattice. Each spin is a one-component object that can point either up or down. The ferromagnetic Ising interaction is particularly simple: if two neighboring spins are parallel (both up or both down), then there is a negative contribution to the energy. Hence the lowest energy of the entire system will be in the configuration in which all the spins in the system are parallel. The Ising model can be regarded as a crude model of a ferromagnet if we think of the classical spins as representing the constituent microscopic moments constituting the ferromagnet. Studies of the Ising model reveal a remarkable feature. If one tunes a control parameter – the temperature T – then one finds that at a certain critical value T_c, spins remarkably far apart have orientations that are strongly correlated. Such correlations do not lend themselves to a ready explanation because normally in physics modeling, the conventional wisdom holds that "you get out what you put in." In the case of the Ising model, we "put in" an interaction that extends a finite distance, yet, almost magically, we "get out" a correlation that spreads an *infinite* distance. How does this happen?

Our intuition tells us that the correlation $C(r)$ between subunits separated by a distance r should decay exponentially with r – for the same reason that the value of money stored in one's mattress decays exponentially with time (each year it loses

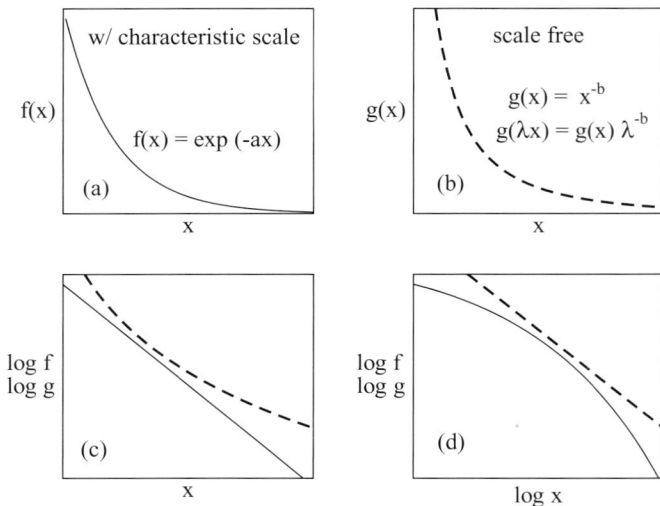

Figure 2.1 (a) Functions that have well-defined characteristic scales (i.e., when all moments are finite) do not have scale-free properties. (b) In contrast, power law functions have diverging moments and scale-free properties. If the scale on the x axis is changed, then the zoomed or dilated function remains unchanged, except for a scaling factor. Scale invariance characterizes fractal phenomena, and power laws constitute the signature of scale invariance. (c) Whereas exponentials appear straight on semilog plots, (d) power laws appear as straight lines on double-log plots.

a constant fraction of its worth, assuming fixed inflation). Thus $C(r) \sim e^{-r/\xi}$, where ξ is termed the correlation length – the characteristic length scale above which the correlation function is negligibly small.[1] Experiments and calculations on mathematical models confirm that correlations do indeed decay exponentially so long as the system is not exactly *at* its critical point, where the rapid exponential decay turns into a long-range power law decay of the form $C(r) \sim r^{-\eta}$ in two dimensions, where η is called a *critical exponent* [357]. (In d dimensions, it decays as $C(r) \sim r^{2-d-\eta}$.) If correlations decay with a power law form, we say that the system is *scale free* because there is no characteristic scale associated with a simple power law (Figure 2.1).

Critical exponents, such as η, are found empirically to depend most strongly on the system dimension and on the general symmetry properties of the constituent subunits, and not on other details of the system under investigation. We understand power law decay as arising primarily from the multiplicity of interaction paths that connect two spins in dimensions larger than one [360]. Exact enumeration methods

---

[1] The symbol $\sim$ is part of Bachmann-Landau notation (see Section 4.3).

take into account exactly the contributions of such paths – up to a maximum length that depends on the computational resources available (and the patience of the investigator). To obtain quantitative results, the hierarchy of exact results is extrapolated to infinite order. In some sense, although the correlation along each path *decreases* exponentially with the length of the path, the number of such paths *increases* exponentially. The *gently decaying* power law correlation emerges as the victor in this competition between the two warring exponential effects.

Cyril Domb, Michael Fisher, Walter Marshall, and their colleagues pioneered exact enumeration approaches in the 1970s. Among the many results emerging from their efforts was a calculation of the scaling equation of state in a magnetic field. A remarkable feature of such work is that the system studied does not perfectly mirror the conditions of the model definition, yet the scaling equation of state measured conforms almost perfectly to the calculated result. For example, if the interactions are antiferromagnetic instead of ferromagnetic in the Ising model, then the magnetization no longer parameterizes the phases transition – but still, the critical behavior is the same [95, 120]. How can that be?

The quick answer is one word: *universality*. The word *universality* connotes the fact that quite disparate systems behave in a remarkably similar fashion near their respective critical points simply because, near their critical points, what matters most is not the details of the microscopic interactions but rather the nature of the "paths along which order is propagated" [361]. Some of the experiments in question are essentially two-dimensional, and the component magnetic moments have sufficient anisotropy in their interactions such that, when acting collectively, they behave as one-dimensional classical spins. Hence the experimental system mirrors a two-dimensional Ising model!

At one time, it was imagined that the scale-free case was relevant to only a fairly narrow slice of physical phenomena [357]. However, the range of systems that apparently display power law correlations has increased dramatically in recent years, ranging from granular materials [152], base pair correlations in DNA [362, 392], lung inflation [362], and interbeat intervals of the human heart [362, 391] to complex systems involving large numbers of interacting subunits that display "free will," such as govern city growth economics [222, 362]. Similar behavior is found in integer numbers [215], and even music [166]. The key concept here involves *fractals* [220] that have scale-invariant properties [64, 65, 114, 359, 362]. Even complex social phenomena, such as teen birth statistics, show fractal scaling [407]. Similarly, the function and internal structures of addiction (e.g., drugs, gambling) and motivation may have a scale-free organization [74].

A relevant example of a fractal interface appears in ecology, in the problem of the territory covered by N diffusing particles [199]. The territory initially grows with the shape of a disk with a relatively smooth surface until it reaches a certain size, at which point the surface becomes increasingly rough (see also [16, 385]). This phenomenon may have been observed by Skellam [349], who plotted contours delineating the advance of the muskrat population and noted that, initially, the contours were smooth but, at later times, they became rough (see Section 5.1). Similarly, foraging trajectories are not smooth but rather appear fractal when analyzed at specific scales [161, 335].

The principle of universality seems to be reflected in the empirical fact that diverse systems can have remarkably similar critical exponents – perhaps the interaction paths between the constituent subunits dominate the observed cooperative behavior more than the detailed properties of the subunits themselves. Even more drastically, perhaps the interactions themselves are not as important as the paths they take.

In the context of the foraging problem, the question concerns whether diverse organisms in a variety of environments move in similar ways. The best-known random walk is Brownian motion and, for a long time, Brownian motion was assumed to describe the diffusion of biological organisms due to the central limit theorem. Indeed, some bacteria [183] may indeed be described by Brownian motion or similar behavior. However, many other organisms do not seem to move according to Brownian motion (see Chapter 6). Instead, they are better described by fat-tailed dispersal kernels. It was in this context that Lévy walks and flights came to play an important role in the study of biological foraging.

2.5 The extraordinary success of limiting models

A common criticism of some models used by physicists studying animal movement is that they ignore basic facts known to ecology and neuroscience, and thus are too simple or naive to have any value. But, one can also argue that such models are useful not in spite of but rather precisely because of the fact that they neglect specific biological mechanisms and other microscopic details. This last statement may surprise readers not familiar with the long tradition in statistical physics of dealing with limiting models [366]. Hence it is worthwhile to discuss this point briefly here.

We already gave one example earlier in this chapter of an apparently simple model that met with unexpected success, namely, the Ising model. Another well-known limiting model is the self-avoiding walk used to describe the behavior of real polymers [100]. Both models neglect the microscopic details of the systems they intend to describe. In fact, real magnets and real polymers bear little resemblance to the Ising model and self-avoiding walks, respectively. Nevertheless, they capture some of the most important aspects of the behavior of the real systems.

Note that it is irrelevant whether the Ising model or the self-avoiding walk is right or wrong because both are wrong by construction. Instead, the relevant question concerns how and why limiting models can mimic real physical systems in crucial ways. There is no doubt that both models have been extremely useful in advancing physics, mainly by capturing the essential features of actual phenomena and thereby helping to shed light on the underlying physical principles and dynamical mechanisms.

Limiting random walk models of biological movement studied by physicists often neglect the specific details of the studied animal [35]. For instance, Lévy walk models of foraging almost completely ignore the role of the brain and nervous system, except to the extent that they are needed to detect nearby prey. One could thus ask whether it makes sense to model a foraging animal as a "brainless" random walker (i.e., a walker with no internal states).

Our view is a bold "yes," because a brainless walker represents a first-order approximation of a perceptive – or even conscious – animal with limited cognitive capabilities. Obviously, we do not deny that some animals have sophisticated nervous systems, brains, etc. Rather, we deliberately choose to study models that minimize the role of cognition in order to focus better on other aspects such as the statistical properties of the movement.

Having analyzed and understood the limiting models, one can then add successive layers of complexity to take into account the roles of the nervous system. Brains can, in principle, always be added or grafted onto "brainless" models.

Although cognition, consciousness, and other functions of the brain may be important in the reactive aspect of biological diffusion-reaction processes (e.g., mating), a brain is completely unnecessary for diffusion. For instance, fungal spores, plant seeds, and pollen are brainless, yet they manage to diffuse adequately.

In contrast to limiting models, such Lévy walks, the more biologically motivated biased correlated random walk (CRW) paradigm attempts to link explicitly information-processing capability with tortuosity. Such biologically specific models do not serve the same purpose as limiting models. Instead, the two types of models play complementary roles and shed light on different aspects of movement phenomena [21, 23]. Hence the two approaches are not mutually exclusive. For example, Pasternak *et al.* [274] have proposed a model named *Lévy taxis* that combines the most important aspects of Lévy walks with those of CRWs. Lévy taxis is a correlated Lévy walk [116].

Nevertheless, the advantages conferred by limiting models, compared to more specific models, often are overlooked or misunderstood. Lévy walks, for example, are not accepted in some quarters because they ignore the underlying specific biological mechanisms (but see Section 14.2). Nevertheless, Lévy walk models have led to significant theoretical and experimental advances, and improvements in data analysis. Moreover, the relative lack of specificity of limiting models means that they can describe not only animals with brains but also pollen and viruses that entirely lack self-propulsion or nervous systems. We will return to these issues in greater detail in Part III.

3

Random walks and Lévy flights

3.1 Central limit theorems

The Gaussian (or "Normal") distribution correctly describes an amazing variety of phenomena, not at the "microscopic" level of single events but rather at the "macroscopic" or statistical level (see Figure 1.3). These bell-shaped curves appear in nature ubiquitously due to the wide applicability of the central limit theorem, which states that the distribution for the sum of a large number of statistically independent and identically distributed random variables that have a finite variance converges to a Gaussian.

Khinchin [181], in his renowned book on the foundation of equilibrium statistical mechanics, based his arguments on (1) ergodic theory and (2) the central limit theorem. The necessary and sufficient conditions for the theorem to hold are sweeping, which explains the ubiquitous, but not universal (e.g., see [412]), finding of Gaussian distributions. Even the Maxwell-Boltzmann distribution of velocities of gas particles corresponds to a special case of the Gaussian distribution (for the velocity vectors of the particles).

We do not include here a proof of the central limit theorem due to its wide availability elsewhere. Given the fundamental importance of the theorem, however, we briefly outline the main ideas involved. Consider the sum S of N independent and identically distributed random variables with zero mean and unit variance. Recall that the probability density function of S equals the Fourier convolution of the N individual probability density functions for the N random variables. The characteristic function is defined as the Fourier transform of the probability density function. By the convolution theorem, the characteristic function of S equals the product of the characteristic functions of the individual random variables. One can show that for large N, the characteristic function of the sum S converges to a Gaussian. The Fourier transform of a Gaussian is itself another Gaussian; hence, the probability density function for S itself converges to a Gaussian. The theorem guarantees that the variance of this Gaussian will grow asymptotically linearly in N.

For a long time, it was assumed that the sum of independent and identically distributed variables always converges to a Gaussian distribution. Only after the 1930s did the scientific community begin to take note of the seminal work of Paul Lévy [209], which showed that the Gaussian distribution is a special case of the more general Lévy stable distribution. One asks, what functional form can the probability density function take for the sum S of independent random variables? The desired functional form must remain invariant under repeated convolutions with the probability density functions of the individual random variables. Equivalently, the characteristic function of S must equal the product of the characteristic functions of the N random variables, leading to the following stability condition: the product of two such characteristic functions must have the same functional form as the original characteristic functions. The Gaussian has this property: the product of two Gaussians is another Gaussian. But other functions of the exponential family have the same property. One can show straightforwardly that the most general stable characteristic function has a stretched exponential functional form, from whose Fourier transform one obtains the expression for the Lévy stable distribution. The most general case, known as the skew Lévy α-stable distribution, is given by

$$\varphi(t) = \exp[itv - |ct|^{\alpha}(1 - i\beta\,\mathrm{sign}(t)\Phi)], \tag{3.1}$$

$$\Phi = \begin{cases} \tan[\alpha\pi/2], & \alpha \neq 1 \\ -(2/\pi)\ln|t|, & \alpha = 1 \end{cases}, \tag{3.2}$$

$$P(S) = \frac{1}{2\pi}\int_{-\infty}^{\infty} dt\, \exp[-itS]\varphi(t), \tag{3.3}$$

where v represents a shift, β is the asymmetry (hence the skewness), and c is a scale.

The Lévy index $\alpha \in (0, 2]$ represents the most important parameter; indeed, it is the key quantity in the study of Lévy walks and Lévy flights. For $\alpha \in (0, 2)$ the probability density function has an asymptotic power law tail with exponent $\mu = \alpha + 1$. Only for three special cases does it lead to closed form expressions: the Gaussian distribution ($\alpha = 2$) with variance $\sigma^2 = 2c^2$ and mean μ (β irrelevant), the Cauchy distribution ($\alpha = 1$, $\beta = 0$), and the Lévy distribution with closed form ($\alpha = 1/2$).

3.2 Normal diffusion and Brownian motion

At the foundation of Brownian motion is the central limit theorem. Indeed, the application of a large number N of independent and identically distributed random impulses to a particle in a viscous fluid will produce a Gaussian distribution for the displacement of the particle. Diffusion governed by Gaussian probability density functions whose variance grows linearly in time is known as *normal* diffusion. In anticipation of the following discussion of *anomalous* diffusion, we briefly mention here three formalisms for treating normal diffusion based on the Langevin, Fokker-Planck, and Master equations.

Langevin equations for a random walker in a potential subject to a stochastic force represent one approach. The Langevin equation is a stochastic differential equation for the velocity in terms of a noise, typically a Wiener process. Numerical integration of Langevin equations leads to the trajectory of the walker.

Often we do not need to know individual trajectories to calculate quantities of interest but only the probability density function for the walker's position. By considering an ensemble of (noninteracting) random walkers, the Langevin equation leads to a Fokker-Planck equation for the probability density function of a walker's position.[1] The best-known Fokker-Planck equation corresponds to free random walkers subject to Wiener noise, in which case the probability density function satisfies the linear diffusion equation, also known as the heat equation,

$$\frac{\partial}{\partial t} P = D \frac{\partial^2}{\partial x^2} P, \tag{3.4}$$

where D denotes the diffusion constant. For the initial condition corresponding to random walkers starting from the origin at time $t = 0$,

$$P_0(x) = P(x, 0) = \delta(x),$$

where $\delta(x)$ is the Dirac delta function, the solution is given by,

$$P(x, t) = \mathcal{N}(t) \exp\left[-\frac{x^2}{4Dt}\right], \tag{3.5}$$

$$\mathcal{N}(t) = \frac{1}{\sqrt{4\pi Dt}}. \tag{3.6}$$

This probability density function $P(x, t)$ is also known as the propagator because $P(x - x_0, t - t_0)dx$ is the probability for finding the random walker in the position interval $[x, x + dx]$ at time t if, at an earlier time t_0, the walker was at position x_0. Note that because the propagator is of the form

$$P(x, t) = \frac{F(x^2/t)}{\sqrt{t}}, \tag{3.7}$$

---

[1] Interactions between random walkers, on the other hand, lead to a hydrodynamic description in terms of convective-diffusive or similar equations, a topic that lies beyond the scope of this book [60, 193].

therefore,

$$\langle x^2 \rangle \sim t$$

$$\langle x^{2n} \rangle \sim t^n.$$

Hence the moments grow in time according to

$$\langle x^{2n} \rangle \sim \langle x^2 \rangle^n. \tag{3.8}$$

In other words, the moments do not grow independently. Moreover, it follows that $\langle x^2 \rangle / t$ is constant. The constant of proportionality can be obtained from Equation (3.5). The mean squared displacement thus grows according to

$$\langle x^2 \rangle = 2Dt. \tag{3.9}$$

The deeper reason for this scaling relation can be intuited from the following fact: the diffusion equation has linear differential operators in the proportion of one time derivative for two space derivatives (even in the higher-dimensional case $\partial_t P = D \nabla^2 P$). Based on this line of reasoning, it is not surprising that if one uses fractional derivatives in time or space, or both, then the mean squared displacement need no longer grow linearly in time. Instead, superlinear and sublinear growth are possible, a point we discuss further later.

A third approach to normal diffusion comes from the discrete analogue of the Fokker-Planck equation for continuous systems. Chapman-Kolmogorov equations, for example, can be used to describe a random walk on a lattice. Similarly, the (Pauli) Master equation describing a random walker jumping between sites on a lattice is given by

$$\frac{d}{dt} P_k = \sum_{\ell} (W_{k,\ell} P_\ell - W_{\ell,k} P_k), \tag{3.10}$$

where $P_k(t)$ represents the probability of being in state k at time t and $W_{k,\ell}$ are the transition rates to go from site ℓ to site k. In the continuum limit, one recovers a Fokker-Planck equation. In fact, all three approaches – the Master equation, Fokker-Planck equation, and Langevin equation – are formally equivalent. They all describe Markovian walks. Normal diffusion arises in the long time limit provided that (1) there is stochasticity (nondeterministic kinematics) and that (2) the transition rates have a range with finite variance.

The simplest case of normal diffusion is the archetypal random walk resulting from nearest-neighbor "interactions" $W_{k,l} = (1/2)\delta_{k-l}^{\pm 1}$, where δ_i^j denotes the Kronecker delta function (with the superscript used merely for convenience). The introduction of short-range memory effects does not lead to significant changes. Short-range (i.e., Markovian) temporal correlations in the noise appearing in the

Langevin equation, for example, can lead to ballistic behavior at short times and diffusive behavior at large times. An example of such correlated random walks [201, 396] is small-angle scattering in two or three dimensions: successive random walk step vectors differ only by some (small) angle, so the small time behavior is ballistic, but for large time intervals, the velocity vectors become independent, leading to diffusive behavior. Similarly, the telegrapher's equation shows coherent, wavelike (ballistic) behavior at short times and diffusive behavior at long times [176] (and a front beyond which the probability is zero due to the finite speed of wave propagation). On the other hand, if the memory has long-range effects, then the behavior may never converge to normal diffusion, even in the infinite time limit (e.g., fractional Brownian motion; see later discussion).

3.3 Anomalous diffusion

Subdiffusion and superdiffusion

Diffusion not described by normal diffusion in the long time limit has become known as *anomalous* [149]. One usually defines the Hurst [16, 164] exponent to quantify how the mean squared displacement $\langle x^2 \rangle \sim t^{2H}$ grows with time t (see Figures 3.1 and 3.2). Superdiffusion corresponds to the case where the mean squared displacement grows superlinearly [353] in time ($H > 1/2$), whereas subdiffusion leads to sublinear [353] scaling in time ($H < 1/2$). Subdiffusion can arise through diverging pausing times in continuous time random walks (see below) as well as via temporal correlations. Single-file diffusion [216] is a well-known example ($H = 1/4$). In biological motion, subdiffusion can arise via long-range temporal correlations, e.g., between turning angles in multidimensional correlated random walk models [147]. The main focus of this book, however, is superdiffusion, with Lévy flights and walks being the best-known example.

Anomalous diffusion with $H = 1/2$

Anomalous diffusion is most commonly defined as $H \neq 1/2$. However, in a weaker sense, anomalous diffusion can also include the case $H = 1/2$, but with non-Gaussian probability density functions. For example, the logarithm of stock prices follows what appears to be a random walk, with $H = 1/2$ and financial returns appearing (misleadingly) uncorrelated. In fact, the absolute returns – the absolute value (or local variance [316]) of the random walk step sizes [222] – possess long-range power law correlations. The relevant probability density functions are not Gaussian because of these long-range memory effects. Moreover, there are *fat tails* in the distribution of financial returns. The presence of long-range power law

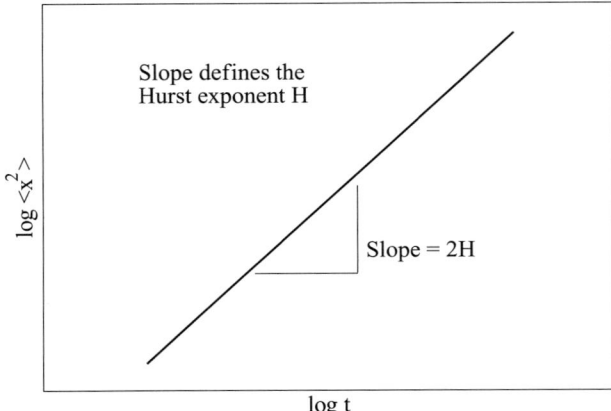

Figure 3.1 The Hurst exponent equals half the slope in the double-log plot of the mean squared displacement versus time for a random walk. Alternatively, it equals the slope of the double-log plot of the root mean squared displacement.

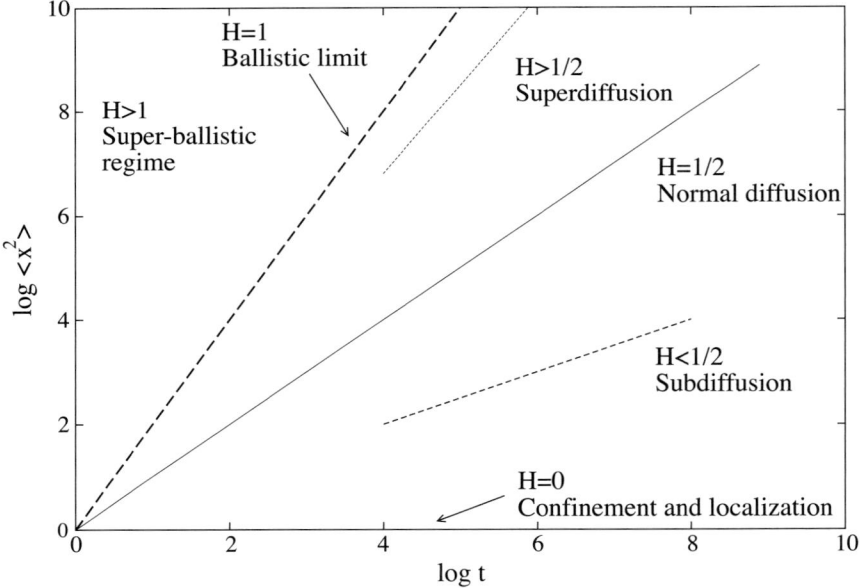

Figure 3.2 The Hurst exponent quantifies key aspects of diffusive processes. The usual uncorrelated Brownian random walks have $H = 1/2$ due to the central limit theorem. Subdiffusive processes have $H < 1/2$. Localized and confined random walkers have vanishing H. Superdiffusion corresponds to $H > 1/2$, with $H = 1$ being the ballistic limit. Superballistic motion ($H > 1$), not treated in this book, corresponds to accelerating particles. Lévy flights and walks are superdiffusive.

correlations and fat tails in the probability density functions is inconsistent with normal diffusion.

Generalized Hurst exponents

The variance (or the second moment) is not always sufficient to quantify the full behavior of probability density functions. A complete description may require an infinite number of moments. The generalized Hurst exponents $H(q)$ quantify the scaling [8] of other moments of the position of the random walker:

$$\langle |x|^q \rangle \sim t^{qH(q)}. \tag{3.11}$$

For example, Equations (3.8) and (3.9) imply $H(q) = H = 1/2$ for normal diffusion and Brownian motion.

Monofractal walks have a single exponent $H(q) = H$, whereas multifractal walks have different (i.e., nonunique) $H(q)$ for different q. One can interpret the difference between them as follows. Monofractals have a single scaling exponent (e.g., fractal dimension). Multifractal walks, in contrast, can be interpreted as consisting of many monofractal walks, each with its own scaling exponent. The probability density function for a multifractal walker starting at the origin changes its shape with time, whereas it only widens (but retains its basic shape) for a monofractal walk.

However, note that monofractality does not imply normal diffusion. Fractional Brownian motion, described by the propagator

$$P(x, t) = \mathcal{N}(t) \exp\left[-\frac{x^2}{4Dt^{2H}}\right], \tag{3.12}$$

$$\mathcal{N}(t) = \frac{1}{\sqrt{4\pi Dt^{2H}}}, \quad 0 < H < 1, \quad H \neq 1/2, \tag{3.13}$$

is monofractal because $H(q) = H$. Yet $H \neq 1/2$, so fractional Brownian motion does not produce normal diffusion. It is an example of monofractal anomalous diffusion that is caused by long-range power law autocorrelations in the underlying noise that drives the random walk. Such "memory" effects constitute one of two mechanisms for disrupting the convergence to normal diffusion implied by the central limit theorem. Long-range memory effects violate the condition of independent random variables.[2]

---

[2] The usual view holds that autocorrelation implies the absence of statistical independence. Actually, this is only one way to account for nonzero autocorrelation. Figueiredo *et al.* [118] show that an independent but nonidentically distributed stochastic process can also lead to a correlated process.

The second mechanism for disrupting convergence to Brownian motion in the long time limit is via power law tailed distributions in the random walk steps (i.e., power law distributed noise rather than Wiener or similar noise). For specific values of the power law exponent, the random variables can have diverging variance, leading to Lévy processes. The classic example, Lévy flights (discussed in Section 3.4), are monofractal. In contrast, the propagator for Lévy walks (also discussed later) has a more complicated behavior at short times, even though the long time behavior becomes similar to that of Lévy flights. We refer the interested reader to the review articles by Metzler and Klafter [241, 243] on anomalous diffusion.

Mathematical formalisms for anomalous diffusion

In analogy with normal diffusion, there are three formally equivalent approaches to anomalous diffusion [136]. The most well known approach uses the formalism of continuous time random walks (CTRWs) [250, 326, 327, 343]. CTRWs were proposed by Montroll and Weiss [250] in 1965. Later, the formalism was shown to be able to handle and explain anomalous diffusion in physical systems. CTRWs allow a generalization of the Wiener processes underlying Langevin equations. The main ingredient in this generalization involves allowing more general distributions of the random walk *jump sizes* and *pausing times*.

A CTRW allows a random walk step j to take a time τ_j and have size ℓ_j. The standard random walk corresponds to the special case $\tau = 1$ and $\ell = \pm 1$. Anomalous diffusion can arise via diverging moments for ℓ or τ. Let $\psi(\ell, \tau)$ denote the joint probability density function for a step to take a duration τ and have length ℓ. Then the jump length and jump time probability density functions satisfy

$$\lambda(\ell) = \int_0^\infty d\tau \, \psi(\ell, \tau) \tag{3.14}$$

$$w(\tau) = \int_{-\infty}^\infty d\ell \, \psi(\ell, \tau). \tag{3.15}$$

The mean pausing time T and second moment of the jump size are then given by

$$T \equiv \int_0^\infty d\tau \, \tau w(\tau) \tag{3.16}$$

$$\sigma^2 \equiv \int_{-\infty}^\infty d\ell \, \ell^2 \lambda(\ell). \tag{3.17}$$

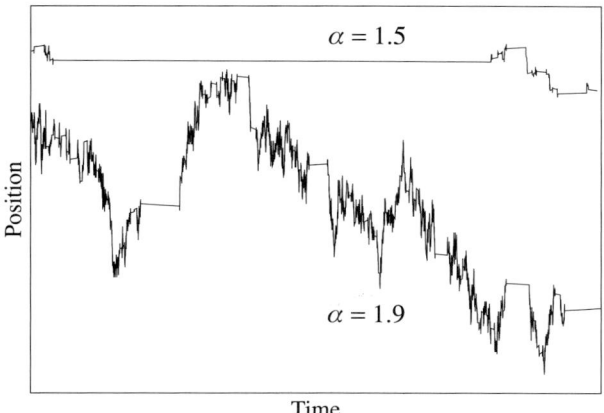

Figure 3.3 Continuous time random walks with power law tailed distributions of pausing times, $P(\tau) \sim \tau^{-\alpha}$. Note how there are long pauses or rests for the lower value of α. For $\alpha \leq 1$, the pausing time distribution has infinite mean pausing time. The result of such ultralong pausing periods is subdiffusion.

A finite value of T guarantees that the number of jumps remains linear in t. Similarly, a finite value of σ^2 ensures that the central limit theorem holds. However, anomalous behavior arises if either value diverges.

If T diverges, then the number of jumps will scale sublinearly with time t, leading to subdiffusion (Figure 3.3). Diverging pausing or "trapping" times for random walkers can lead to ergodicity breaking so that the system can fall into an out-of-equilibrium regime, with aging phenomena [226]. *Aging* in physical systems refers to qualitative changes that occur due to incremental accumulation of memory or information about the past over a long period of time (e.g., structural damage in aging materials due to fatigue [153]). In contrast to systems without aging, such systems typically have nonexponential relaxation. There are many other interesting features of subdiffusion [80, 196, 216, 325, 416], but their discussion lies beyond the scope of this book.

We present here a scaling argument for the case $w(\tau) \sim \tau^{-(\alpha+1)}$, when $T \to \infty$ ($\alpha < 1$). Let t_N denote the time for jump number N. Then, from Lévy's generalized central limit theorem, we know that

$$t_N = \sum_{i}^{N} \tau_i \sim N^{1/\alpha}, \tag{3.18}$$

which we can invert to obtain a definition of the natural *operational time* for the random walk process, which is not the actual time but is rather (in some arbitrary system of units) equal to the number of steps:

$$\text{operational time} = N \sim t_N^{\alpha}. \tag{3.19}$$

If σ^2 is finite, the central limit theorem applies to the operational time,

$$\langle x^2 \rangle \sim N, \tag{3.20}$$

from which we obtain the desired scaling relation [343]:

$$\langle x^2 \rangle \sim t^\alpha. \tag{3.21}$$

Therefore the mean squared displacement grows sublinearly in time. This is the well-known Scher-Montroll CTRW with power law distributed pausing times. In contrast, if σ^2 diverges, then one obtains superdiffusion and Lévy flights, described in Section 3.4.

If the jump times are independent (or decoupled) from the jump lengths, then such CTRWs are separable, examples of which include Lévy flights and the Scher-Montroll CTRW discussed above. In contrast, Lévy walks (see below) and continuous-time Weierstrass flights [194] are nonseparable. Weierstrass functions are everywhere nondifferentiable but continuous functions. Historically it was one of the first fractal curves, although the term *fractal* did not exist during Weierstrass's life.

Of great importance is the fact that it is possible to obtain analytical (sometimes even exact) solutions using the CTRW formalism. Hence experimental first-passage time analysis [112], mean squared displacements, etc., can be meaningfully compared with theoretical estimates. We thus briefly discuss the analytical treatment of CTRWs.

Consider the position $x(t)$ of a random walker expressed as the sum of the individual steps ℓ_n taken,

$$x(t) = \sum_{n=1}^{N(t)} \ell_n, \tag{3.22}$$

where $N(t)$ denotes the total number of jumps taken until time t. Let $\mathcal{P}(N, t)$ denote the probability distribution for taking $N(t)$ total steps until time t, and let $A * B$ indicate a Fourier or Laplace convolution in the time or space domain, respectively,

$$(A * B)(t) = \int_0^t dt' A(t') B(t - t')$$

$$(A * B)(x) = \int_{-\infty}^{\infty} dx' A(x') B(x - x').$$

Moreover, let f^{*N} indicate convolution to the Nth power, i.e., N convolutions of f with itself. Then, for such separable CTRWs, the probability density function for

$x(t)$ can be written as

$$P(x, t) = \sum_{N=0}^{\infty} \left[\lambda^{*N}(x) \mathcal{P}(N, t) \right] \tag{3.23}$$

$$= \sum_{N=0}^{\infty} \left[\lambda^{*N}(x) \left(w^{*N}(t) * \int_t^{\infty} d\tau\, w(\tau) \right) \right], \tag{3.24}$$

where λ and w obey Equations (3.14) and (3.15). Note here that in addition to N "pauses," we must explicitly consider the probability of not completing an event before time t, which is given by $\int_t^{\infty} w(\tau)d\tau$.

Taking the Fourier transform of P in x and the Laplace transform in t, we arrive at the Montroll-Weiss equation:

$$P(k, s) = \sum_{N=0}^{\infty} \left[\hat{\lambda}^N(k) \tilde{w}^N(s) \left(\frac{1 - \tilde{w}(s)}{s} \right) \right] \tag{3.25}$$

$$= \left(\frac{1 - \tilde{w}(s)}{s} \right) \left(\frac{1}{1 - \tilde{w}(s)\hat{\lambda}(k)} \right). \tag{3.26}$$

Hence, for arbitrary jump length and pausing time distributions, an analytical solution can be obtained in the Fourier-Laplace domain. Usually, all that matters is the behavior near the origin in the Fourier-Laplace domain, i.e., the long time behavior of the *tail* of the propagator. For example, series expansions (keeping the lower terms) can be used to calculate the inverse transforms to obtain the desired quantities.

A second formalism for dealing with anomalous diffusion is the generalization of Fokker-Planck equations with fractional time or space derivatives. For a full treatment of the definition of Riemann-Liouville, Riesz, and Caputo fractional derivatives, we refer the reader to the literature on the subject. A comprehensive treatment of anomalous diffusion and transport can be found in the book by Radons *et al.* [291]. We briefly outline some of the main ideas, however, to give the reader a flavor for the subject.

Consider the linear differential operator $\partial/\partial x$, whose Fourier transform is (ik). Note that the differentiation operator is a Fourier multiplier. The Laplacian operator ∇^2 becomes $-k^2$ under Fourier transformation. Exploiting this correspondence, we can define a fractional Laplacian of order $\alpha/2$ as the operator whose Fourier transform is $-|k|^{\alpha}$. Taking advantage of such properties, one can obtain any Hurst exponent in the range $0 \leq H \leq 1$ by generalizing the diffusion equation via fractional time and space derivatives [241, 243].

A third formalism for dealing with anomalous diffusion is the generalized master equation [175, 178, 180]. The master equation is necessarily local in time, even though it can have long-range spatial coupling (see Equation (3.10)). The non-Markovian analogue of the master equation is the generalized master equation

$$\frac{d}{dt}P_k = \sum_\ell \int_0^t dt'[W_{k,\ell}(t-t')P_\ell(t') - W_{\ell,k}(t-t')P_k(t')], \qquad (3.27)$$

such that the dynamics at time t at site k depends not only on what is going on at other sites ℓ, but also on what happened at previous times $0 \le t' < t$. In other words, the generalized master equation adds memory effects to the usual master equation.

CTRWs, generalized master equations, and fractional Fokker-Planck diffusion equations are all formally equivalent [136]. There are some issues relating to inversion of Laplace transforms, but such issues are not important if we allow generalized functions, tempered distributions, etc.[3]

3.4 Lévy flights and Lévy walks

Mandelbrot coined the term *Lévy flight* in his book on fractal geometry [220], in the context of Lévy dusts, i.e., fractal dusts generated by the turning points of a Lévy flight. Lévy flights and Lévy walks have found application in the study of diverse systems and phenomena [330, 331, 342]: bulk mediated excursions [87], finance and economics [5, 222, 290, 363], protein folding dynamics [233], kinematics of ions in optical lattices [174], cavity quantum electrodynamics [398], climate [375] and atmospheric physics [376], anomalous spin relaxation [217], soft-mode turbulence [369], dispersive sedimentation [240], optics [279] (e.g., the Lorentz line shape), photonic superdiffusion [4], random lasers (and materials through which photons execute a Lévy walk; see [410] and references therein), metastability [165], knots [251], bioturbation [248], and anomalous diffusion of surfactants [7].

Lévy flights arise when the jump size distribution has a power law tail $\lambda(\ell) \sim \ell^{-\mu}$, leading to diverging variance for $\mu < 3$. The necessary and sufficient conditions of the central limit theorem do not hold in this case. Instead, one finds that the probability density function for the position of the walker converges to a Lévy stable distribution with Lévy index $\alpha = \mu - 1$, with $0 < \alpha \le 2$, with the special case $\alpha = 2$ corresponding to normal diffusion (Figure 3.4).

When $\alpha < 2$, the behavior is superdiffusive, with the limit $\alpha \to 0$ leading to ballistic motion (Figures 3.6 and 3.7). The propagator $P(x,t)$ widens in space superlinearly in time for $\alpha < 2$ (see Equation (3.29)). What matters most, as far as

---

[3] We mention in passing that there have been attempts to frame anomalous diffusion in the context of maximum entropy formalisms, e.g., via nonstandard entropies [414]. However, such ideas have never been applied to the study of foraging successfully.

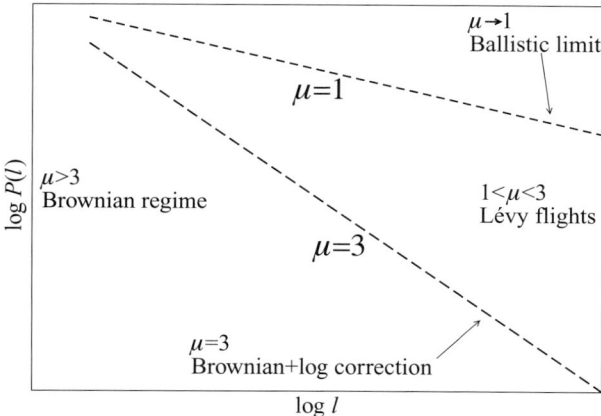

Figure 3.4 For a random walker who takes steps of size ℓ according to a probability density function $P(\ell) \sim \ell^{-\mu}$, the resulting type of diffusion depends on the value of μ. For $\mu > 3$, the central limit theorem guarantees convergence to normal diffusion. The ballistic limit corresponds to $\mu \to 1$. For $\mu \leq 1$, one cannot normalize the distribution P. Intermediate values $1 < \mu < 3$ result in superdiffusive Lévy flights.

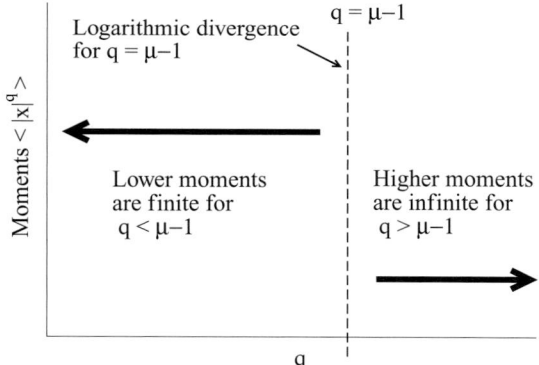

Figure 3.5 For a power law tailed distribution $P(\ell) \sim \ell^{-\mu}$ of jump or step sizes ℓ_j, higher moments do not exist; i.e., they are infinite. Specifically, the moment of order $\mu - 1$ diverges logarithmically with the upper cutoff, and all higher moments diverge as some power of the upper cutoff. Lower moments remain finite. Diverging moments are a consequence of the scale invariance properties: scale-free systems cannot have well-defined characteristic scales.

diffusion is concerned, are the tails of the propagator; i.e., the small flights play a negligible role in many ways [106]. Notice that for $\alpha < 2$, one cannot define the mean squared displacement because it diverges. Instead, one can study moments of order lower than α because they do not diverge (Figure 3.5). In this way, one can define width, such as half widths at half maximum, and show that a *pseudo* mean squared displacement grows as $\sim t^{1/\alpha}$ for Lévy flights. Hence a unique Hurst exponent

$$H = \frac{1}{\alpha} = \frac{1}{\mu - 1} \tag{3.28}$$

characterizes the behavior of Lévy flights. This scaling relation follows from the form of the propagator, which is a Lévy stable distribution. For the case of a Lévy flight with zero drift velocity and symmetric noise, the propagator has the general form

$$P(x, t) = \mathcal{N}(t) F\left(-\frac{|x|}{t^{1/\alpha}}\right), \tag{3.29}$$

where $\mathcal{N}$ is the normalization. Hence the moments satisfy

$$\langle |x|^q \rangle^{1/q} \sim t^{1/\alpha}, \quad q < \alpha, \tag{3.30}$$

from which Equation (3.28) follows immediately.

The master equation for Lévy flights can be obtained by introducing algebraically decaying long-range interactions. Specifically, if the second moment of the hop length distribution diverges, the central limit theorem will not apply, and instead, one will obtain Lévy behavior. There is no need for a generalized master equation such as (3.27) because Lévy flights are Markovian.

A fractional diffusion equation [7, 9, 244, 271] that describes Lévy flights can be obtained from the functional form of the propagator: one power of t is associated with α powers of x. Indeed, Lévy flights bear a relation to generalized diffusion equations having fractional powers of Laplacians [241, 243, 317, 389]. Let the fractional Laplacian $|\nabla|^\alpha \equiv -(-\nabla^2)^{\alpha/2}$ denote the differential operator whose Fourier transform is $-|k|^\alpha$. For $\alpha = 2$ this is the standard Laplacian. Then the fractional diffusion equation,

$$\frac{\partial}{\partial t} P = D_\alpha |\nabla|^\alpha P, \tag{3.31}$$

where D_α is a generalized diffusion coefficient, will have solutions of the form given by Equation (3.29) for $0 < \alpha \leq 2$. By generalizing the Laplacian operator in reaction-diffusion systems using fractional Laplacians in this way, one can introduce superdiffusion into the models. Similarly, diffusion equations with fractional time derivatives allow subdiffusion. Dubkov and Spagnolo [102] and Denisov *et al.* [98] have derived and solved the fractional Fokker-Planck equation [102, 167] for Lévy flights in the steady state for several potential wells. Moreover, superdiffusive reaction-diffusion systems can display remarkably rich behavior [259]. Baeumer *et al.* [13] have established the connection between these anomalous (i.e., superdiffusive) reaction-diffusion systems and corresponding integro-difference equations. Brockmann and Hufnagel have shown [54] that wave fronts in a fractional generalization of reaction-diffusion systems propagate with constant velocity even for superdiffusive particles, for fluctuations involving a finite number of particles per volume.

Although proposed originally in low-dimensional settings, Lévy flights have been generalized to higher dimensions [129, 130]. They have many unusual and interesting properties [55] besides diverging moments and superdiffusion. For example, ordinary diffusion is typically slowed in the presence of spatial disorder. Belik and Brockmann [26] have shown that spatial inhomogeneities can actually facilitate the spread of superdiffusive processes due to the nonlocal nature of the diffusion. Another counterintuitive property is the behavior of Lévy flyers subject to an attractive force. For example, a Lévy flyer in steeper than harmonic potentials can have finite variances. Moreover, the probability density function can become bimodal [78] (i.e., with two peaks rather than a single, central bell-shaped region). Another, more mathematical, property is that the method of images can fail when applied to Lévy flights [77]. Owing to such counterintuitive properties, Lévy flights continue to be the subject of ongoing investigation.

Having discussed Lévy flights, we now turn to Lévy walks [245]. The difference between a Lévy flight and a Lévy walk concerns the velocity. Lévy flight jumps take zero or vanishingly small time, whereas a Lévy walk proceeds at a constant velocity (Figure 3.6). This finite velocity of propagation couples the space and time. Specifically, a Lévy flight is local in time; i.e., it can be described by a Master equation with long-range transition rates in space. Hence it is a Markov process. In contrast, a Lévy walker takes $v\ell_j$ units of time to move a distance of ℓ_j units due to the finite velocity v. Hence the scaling behavior of Lévy walks is more difficult to derive. Lévy walks can be described in terms of a Dirac δ-function coupling of time and space. Let τ_j be the time taken to traverse the distance ℓ_j. Then, using the CTRW formalism, the finite velocity can be expressed via,

$$\psi(\ell, \tau) = \psi(v\tau, \tau) \tag{3.32}$$

$$\propto w(\tau)\delta(\ell - v\tau). \tag{3.33}$$

Lévy walks correspond to power law tails in this nonseparable distribution:

$$\psi \sim \ell^{-(1+\alpha)}\delta(\ell - v\tau). \tag{3.34}$$

Because $\psi(\ell, \tau)$ is not separable, Lévy walks require a generalized master equation or equivalent formalism for their description. (For more information, see Metzler [239] or the original papers.) The scaling behavior of Lévy walks was derived by Geisel *et al.* [133] and also by Klafter *et al.* [185].

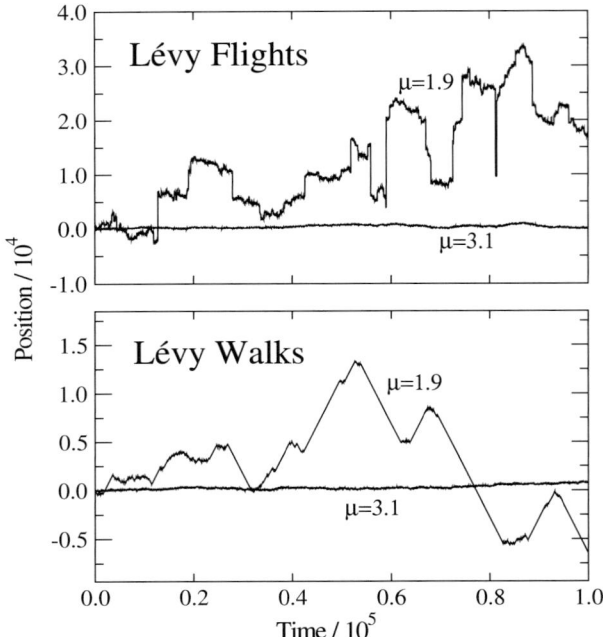

Figure 3.6 Examples of one-dimensional Lévy flights and walks for different values of the power law exponent μ. For $\mu > 3$, the variance of the noise distribution is finite, and hence the random walk is diffusive with Hurst exponent $H = 1/2$. But for $\mu < 3$, the behavior is superdiffusive. The only difference between Lévy flights and walks is that Lévy walkers move with finite velocity, seen by the finite slope in the space-time plots. In contrast, Lévy flyers relocate "instantly."

One can intuitively understand their superdiffusive behavior via an order of magnitude "back-of-the-envelope" scaling argument, which is not rigorous but which is nevertheless instructive. Consider the case $w(\tau) \sim \tau^{-(1+\alpha)}$, which is equivalent to taking $\lambda(\ell) \sim \lambda^{-(1+\alpha)}$ due to the space-time coupling,

$$t \sim N, \quad 1 < \alpha < 2, \tag{3.35}$$

$$x = \sum_i^N \ell_i, \tag{3.36}$$

$$\langle \ell^2 \rangle \sim \int_a^t d\ell \, \ell^2 \ell^{-(1+\alpha)} \sim t^{2-\alpha}. \tag{3.37}$$

Because $\langle \ell^2 \rangle$ is finite but growing, we can expect normal diffusion but with a time-dependent diffusion constant. Specifically, we expect that $\langle x^2 \rangle$ should be proportional to $\langle \ell^2 \rangle$ because the latter represents a characteristic scale of the

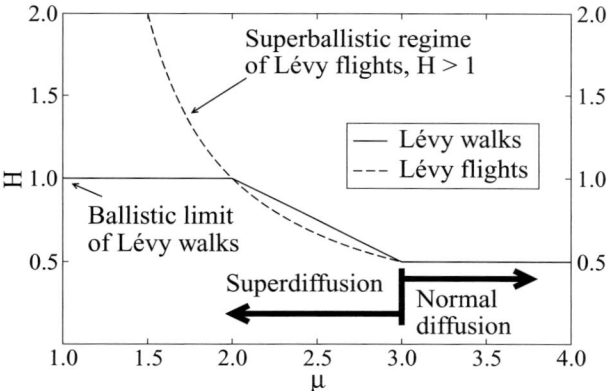

Figure 3.7 The Hurst exponents $H = H(2)$ for the behavior of the mean squared displacement in the long time limit for Lévy walks, as a function of the exponent μ of the power law tail $\sim \ell^{-\mu}$ in the jump size distribution. For $1 < \mu \leq 3$, the corresponding Lévy index is $\alpha = \mu - 1$. Lévy walks are superdiffusive for $\mu < 3$, i.e., for $\alpha < 2$. For comparison, the behavior is also shown for Lévy flights; however, Lévy flights, in fact, have diverging mean squared displacements for $\alpha < 2$, so we cannot define $H = H(2)$. Nevertheless, since Lévy flights are monofractal (see the text), we can still define $H = H(q)$ for $q < \alpha$.

jumps,

$$\langle x^2 \rangle \sim \langle \ell^2 \rangle \, t. \tag{3.38}$$

Substituting the expression for $\langle \ell^2 \rangle$, we arrive at the final result,

$$\langle x^2 \rangle \sim t^{3-\alpha}, \quad 1 < \alpha < 2. \tag{3.39}$$

From this last expression, we find that in the long time limit,

$$H = \frac{3-\alpha}{2} = \frac{4-\mu}{2}, \quad 1 < \alpha < 2 \tag{3.40}$$

for Lévy walks (Figure 3.7). For $\alpha < 1$, the mean jump size itself diverges, so $H = 1$. There are logarithmic corrections for the cases $\mu = 3$ and $\mu = 2$ for the mean squared displacement because the mean (variance) step length diverges logarithmically for $\mu = 2$ ($\mu = 3$). In Chapter 10, we will show that $\mu = 2$ optimizes random searches under certain circumstances.

We thus see that Lévy walks have scaling slightly different from that of Lévy flights, for which $H = 1/\alpha$. Figure 3.7 compares the Hurst exponents of Lévy walks and flights. In the context of foraging, we can use either expression, depending on the nature of the quantities being studied. Indeed, the only difference between walks and flights is the nature of the operational time. If one wishes to study the behavior as a function of the real time t, then one must use the expression for Lévy walks. However, if one wishes to express the scaling in terms of the random walk step number j, then one should use the expression for Lévy flights because j is proportional to the operational time for a Lévy flight.

Metzler [239] has given a helpful explanation for why Lévy walks increasingly look like Lévy flights in the long time limit. At very large times, the behavior of the propagator is dominated by the large jumps so that it slowly converges to a true Lévy stable distribution. At $x = \pm vt$, there will be remnants of Dirac δ-functions that arise from the possibility of an ongoing very large initial jump. Such Dirac δ-functions always arise from ballistic behavior because the ballistic propagator is $P(x, t) = \delta(x - vt)$. Indeed, the $\alpha \to 0$ limit of Lévy walks is ballistic motion, given by

$$P(x, t) = \frac{\delta(x - vt) + \delta(x + vt)}{2}. \tag{3.41}$$

Lévy walks become similar to Lévy flights only after a long period of time. Moreover, the δ-function peaks at the edge of the propagator decay slowly. These properties hint at one of the more interesting properties of Lévy walks, which is also a general property of animate systems: aging [195]. Allegrini *et al.* [3] have shown that Lévy walks are characterized by bilinear scaling and that Lévy walks (but not Lévy flights) undergo aging. This aging occurs because the finite speed of propagation for a Lévy walk postpones to an infinitely long time the exact equivalence (or convergence) of Lévy walks and Lévy flights [3]. Lévy walks are also associated with nonequilibrium phenomena [86, 200, 226, 242, 314, 415].

Although Lévy flights and walks have many features in common, some important differences cannot be neglected. Because infinite velocities are unphysical, Lévy flights are impossible in physical geometric space. Nevertheless, they can occur, for example, along folded polymers such that a particle hopping from one site to another (on a different part of the folded polymer chain) can traverse a very long distance measured along the chain, even though the Euclidean distance may be short.

In contrast, Lévy walks do not violate any physical laws. Specifically, no physically measurable quantity diverges for a Lévy walk in finite time. The infinite variance of the underlying Lévy (or power law) process only becomes an issue at infinite time because a Lévy walker can only traverse a maximum distance of vt in time t. So all moments of the propagator remain finite at all times, unlike for Lévy flights, which have diverging moments. In the context of foraging and biological motion, Lévy walks are more plausible and appropriate.

We end this chapter by warning the reader about diverging quantities. Indeed, the scientific community has been slow to welcome Lévy processes into the mainstream literature, and for valid reasons. Because physically realizable systems can never have diverging moments or other infinite measurable quantities, pure Lévy flights are idealizations that do not find physical realization. All power laws in nature have upper and lower cutoffs. Textbooks rarely mention that even the inverse square law of gravity does not extend to infinite distances because the universe is likely to have a finite age and even gravity cannot propagate faster than the speed of light. Truncations can arise in a number of ways. For example, in the context of physical processes, Chechkin *et al.* [79] have shown that nonlinear dissipation can regularize the velocity distribution of a random walker subject to Lévy noise by introducing an upper cutoff and rendering all moments finite. A Lévy walk (or flight) with an upper cutoff in the power law tail of the flight length distribution is known as a truncated Lévy walk (or flight). Certain kinds of truncated Lévy flights also have analytical descriptions in terms of fractional diffusion equations [354]. The cutoff need not be abrupt, in which case such walks are known as gradually truncated Lévy walks and flights [143, 144]. Truncated walks have Lévy properties at most of the relevant scales, but at "astronomically" large timescales the central limit theorem eventually forces a convergence to normal diffusion [221].

4

The wandering albatross

4.1 Do good theories always come from good data?

According to conventional wisdom concerning the scientific method, good theories come from good experimental data, and bad theories from bad experimental data. Yet the history of the physics of foraging is a remarkable counterexample. To illustrate this, we briefly recount one of the important scientific investigations in the field, published in *Nature* in 1996. The original study of wandering albatrosses [390] inspired dozens of other studies, yet later required correction due to its spurious data.

4.2 Lévy flights of the wandering albatross

The albatross can fly great distances, at exceptional speeds. There are significant differences among species of albatross [402]. Wandering albatrosses in southern Georgia can sustain a speed in excess of 100 km/h by taking advantage of the local wind field [284]. They frequently fly 500 km per day, with an upper limit in the range 750–950 km per day. Phillips *et al.* [284] report that one gray-headed albatross circumnavigated the Southern Ocean in only 46 days. Because of their great mobility and large size, we decided to focus on the albatross (instead of, e.g., the sparrow) in our original study. The foraging strategy of the wandering albatross [403] stands apart from that of other seabirds [401]. Weimerskirch *et al.* [404] studied the distribution of prey encounters for wandering albatrosses and reported results that strongly suggest a foraging strategy that differs from those of most seabirds. Unlike most seabirds, which concentrate in more predictable foraging areas, wandering albatrosses appear to rely on highly dispersed prey.

The unique characteristics of wandering albatrosses rendered them attractive to physicists wanting to study how autonomous behavior arises in complex systems. The implicit assumption was that if we could quantitatively model how animals choose to move, perhaps this advance could help us to better understand decision making and other cognitive processes. There was even the possibility that these studies could shed light on such potentially nebulous concepts as free will [151, 395] (see also Section 14.3).

In the early 1990s, researchers affiliated with the British Antarctic Survey (BAS) collected data at Bird Island. The recording devices, attached to the legs of the seabirds, measured the electrical resistance between two electrodes. When the birds were flying, the dry conditions decreased the electrical current. When they were swimming or submerged in seawater, the electrical current increased. The devices could thus detect and record any flight activity. The logged data set consisted of intervals of wet and dry conditions for each seabird for each trip, and ranged in duration from 77 to 416 hours.

In 1994, Sergey Buldyrev obtained from BAS flight data of a number of wandering albatrosses. Dr. Buldyrev had been in contact with Dr. V. Afanasyev and his collaborators at BAS, who had developed the logging devices. Dr. Buldyrev and the group were at that time studying long-range power law correlations in complex physical and biological systems at Boston University. As a graduate student, one of us (G.M.V.) analyzed the data. The results strongly suggested an inverse square power law for the histogram of flight times, and hence of flight distances, assuming approximately constant velocity. After these results appeared in the scientific literature [390], they motivated many further studies, both theoretical and empirical.

A few years later, however, Dr. Buldyrev noticed that some of the very long flights typical of Lévy flights appeared at the beginning of the time series of the flight data. A careful new analysis (Figures 4.1–4.3) of the data clearly indicated that the power law tails become truncated when the histograms' first and last data points of each time series are excluded. If the first or last points had spurious origins or represented artifacts of the experimental procedure, this would weaken the evidence supporting Lévy flights for wandering albatrosses.

Together with other collaborators, we contacted Dr. Richard Phillips at BAS to reanalyze the data. We needed to check for spurious flights at the beginning of each time series. Dr. Phillips, along with other researchers, used independent data obtained from platform terminal transmitters (PTTs) to cross-check the logger data. The PTT data unambiguously established the spurious nature of the ultralong flights at the beginning of the logger time series [105]. Out of this disappointment came new questions. What had happened, and why?

New investigations were carried out to more fully understand what had transpired and, by 2007, the answers had become clear. The logger data contained long dry periods because each bird had remained for some time under captivity prior to release, and this initial time period caused spurious results that seemed to indicate

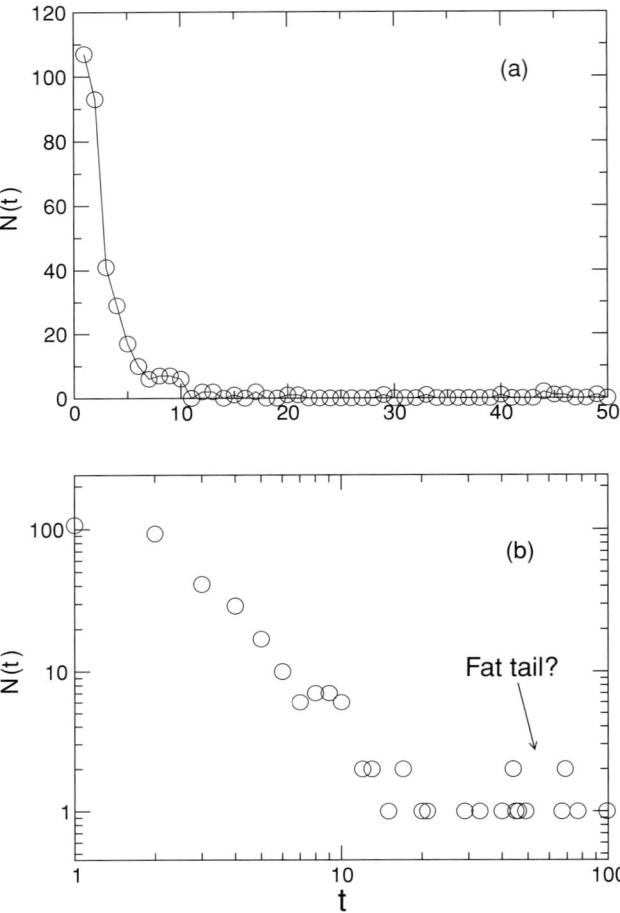

Figure 4.1 Histogram of the flight times (in hours) of wandering albatrosses [390], shown on (a) linear and (b) log-log scales. The histogram decays slowly. The key question is, could this be indicative of a power law tail? This question motivated theoretical research that led to significant advances, yet the original data contained errors, as discussed in the text and shown in Figure 4.3.

extremely long flights. The ultralong flights characteristic of Lévy flights detected in the original study were caused by these spurious data points.

Publication [105] of these corrected results, undertaken in collaboration with BAS researchers as well as with physicists from other institutions, led to controversy. The article, authored by Andrew M. Edwards and coauthored by Richard A. Phillips, Nicholas W. Watkins, Mervyn P. Freeman, Eugene J. Murphy, Vsevolod Afanasyev, Sergey V. Buldyrev, and ourselves, was published in *Nature* in 2007. It

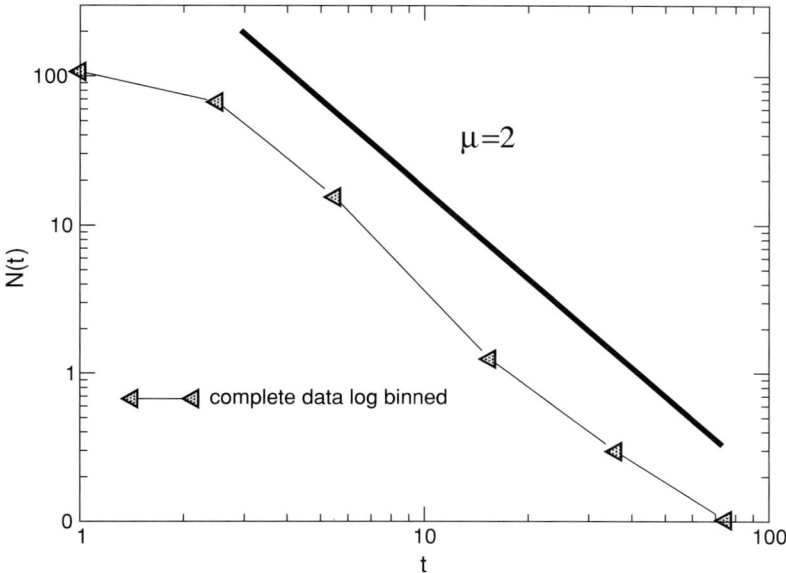

Figure 4.2 Double-log plot of the logarithmically binned histogram of flight times shown in Figure 4.1. The logarithmically binned histogram of the complete data points to Lévy flight behavior. These results, published in 1996 [390], helped to overturn the conventional wisdom about how animals move.

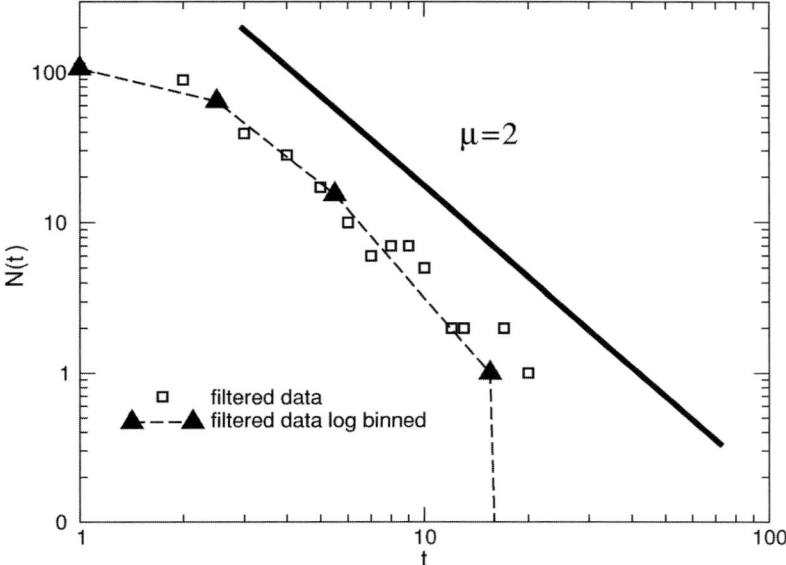

Figure 4.3 Elimination of the first and last flights from each time series leads to an upper cutoff that dramatically diminishes the power law tail seen in Figure 4.2 [105]. The evidence in favor of Lévy flights thus became weaker. Note, however, that a power law with $\mu \approx 2$ is still apparent below the cutoff.

not only corrected the earlier work but also emphasized the importance of rigorous data analysis and statistical inference. The paper also raised questions for a more important reason: it put under renewed scrutiny the very idea of biological Lévy flights and walks. We refer the interested reader to an article by Buchanan [59], who has written clearly and concisely about these issues. Since then, the dust has settled and we now have a clearer understanding. Part II of this book contains a more detailed discussion of the empirical evidence for Lévy flights, while Part III details the theoretical foundation.

4.3 Power laws and Pareto's principle

The preceding explanation also raises a somewhat troubling concern: can one or two data points per time series dramatically skew results? The answer is an unequivocal "yes." The explanation lies in a property of both Lévy processes and power laws, in general.

Power laws have self-affine and scale-free properties (Figure 2.1). A self-similar object is one that is at least approximately identical to a part of itself. A generalization of self-similarity is self-affinity. A self-affine object is one that is at least approximately identical to a linear transformation of itself. In contrast to self-similar objects, the linear transformation of a self-affine object need not be isotropic. *Scale invariance* refers to symmetry under scale transformations. Self-affine and self-similar objects have scale-free properties.

The scale invariance inherent in power laws makes it difficult to treat so-called outliers, or even to define the concept of outlier. In Gaussian (i.e., normal) statistics, an *outlier* refers to an observation that deviates disproportionately from the rest of the data. But when the system or phenomenon has scale-free properties, the rest of the data is statistically similar to the whole data set – hence it will itself contain rescaled outliers, rendering the concept only marginally useful, at best. Indeed, power law tailed distributions may allow extreme "events" that are not rare enough to rule them out (for all practical purposes). An example is the Gutenberg-Richter power law for earthquakes, which predicts that earthquakes of magnitude $M + 1$ will occur 10 times less often than those of magnitude M. Owing to this inherent scale invariance, it makes no sense to define an outlier because arbitrarily large events will occur if one just waits long enough. Only the finite size of the system imposes an upper limit on the largest possible earthquake.

Moreover, 10% of the data points can account for a fat tail consisting of 90% of a power law distribution. Consider, for example, a truncated inverse square power law probability density function with scaling over 2 decades

$$P(x) \propto 1/x^2, \quad 1 < x < 100.$$

The ratio of the numbers of data points falling in the first and second decades of the distribution (i.e., from 1 to 10 and 10 to 100, respectively) is

$$\frac{\int_1^{10} dx/x^2}{\int_{10}^{100} dx/x^2} = \frac{9/10}{9/100} = 10.$$

Thus the second decade contains only approximately 9.1% of the data points, yet it adds an entire order of magnitude to the tail of the distribution. In fact, 90% of the distribution is due to fewer than 10% of the data points.

Even standard deviations and means need to be treated carefully. The mean (i.e., the first moment $\langle x \rangle$) diverges for an idealized inverse square probability density function $P(x) \sim 1/x^2$. Henceforth, the symbol $\sim$ refers to the asymptotic limiting behavior as the relevant quantity goes to infinity or zero. This shorthand, which is part of Bachmann-Landau notation, not only simplifies the notation but often helps to focus attention on scaling. In general, a power law tailed distribution that decays as $P(x) \sim 1/x^\mu$ will have diverging moments $\langle x^q \rangle$ of order $q = \mu - 1$ and higher. For power law tails that decay with $\mu \leq 3$, standard deviations do not exist, hence the need for other quantities to measure width, such as the full width at half maximum.

All fractal phenomena have intrinsic power laws associated with them. The scale-free properties associated with power laws continue to motivate research that attempts to estimate power law exponents from actual data. In contrast, Gaussian, Poisson, and other distributions that have no diverging moments always possess a characteristic scale, allowing a proper, well-defined, physical interpretation of the concept of outliers.

A power law for the flight times of the albatrosses implies ultralong but rare flights. The original data for the wandering albatross contained fewer than 370 data points, obtained from 19 separate trips. The probability density function, estimated via logarithmically binned histograms, appeared to indicate power law scaling over a range of almost two orders of magnitude. For such a distribution, we should expect approximately 90% of the data points to fall within the lower 10% of the range of flight distances. Fewer than 10% of the data points should correspond to points in the second decade of scaling, i.e., to the tail of the distribution. Because there were 19 time series (one for each trip), the first flights account for 19 data points, and the last flights for another 19, totaling 38 data points – crucially, more than 10% of the total.

Thus eliminating only 10% of the data points can deflate 90% of the power law tail. Compare this effect with how the countably few largest earthquakes cause most of the damage. Such counterintuitive behavior can be understood within the larger context of a rule of thumb known as Pareto's law or Pareto's principle. Pareto observed, more than a century ago, that approximately 80% of the wealth in Italy was owned by some 20% of the population [222]. More generally, 70%, 80%, 90%, etc., of the effects can be due to 30%, 20%, 10%, etc., of the causes, respectively. (The choice of distribution determines the particular values.)

One lesson to learn from this episode is that extra care is needed when studying Lévy processes in animal locomotion. The last few years have seen considerable effort go into this endeavor. In addition to the issues discussed in the article by Buchanan [59], mentioned above, another aspect of the research concerns the conclusions. Although the original work on albatrosses required correction, very recent empirical research using rigorous statistical techniques does, in fact, seem to leave open the possibility of Lévy flight foraging patterns for wandering albatrosses. Indeed, Boyer *et al.* [48] and Reynolds [299] subsequently showed that the measured flight distributions for albatrosses do not, in fact, disagree with the hypothesis of a *truncated* Lévy walk. For some other species, the evidence seems to be convincing. Even human movement patterns contain non-Gaussian properties similar to those of Lévy walks (see Part II).

Hence wholesale rejection of the idea of Lévy flights, in reaction to the spurious albatross results, without taking into account the empirical evidence in favor of Lévy flights for many other animal species (reviewed in Part II) is inconsistent with an impartial and objective scientific attitude. In this context, the mature scientific debate on the issue, advanced cautiously and incrementally, has helped end the controversy. In addition to this lesson, however, another issue has to do with the scientific method itself.

4.4 Scientific progress as a random walk

The original study had motivated us to approach the problem from a theoretical perspective. Thinking that albatrosses in fact execute Lévy flights, we sought a plausible explanation for this fact. Our efforts resulted in an unexpected discovery: a Lévy flight process with an inverse square power law of flight distances optimizes the search of randomly but sparely located revisitable targets [393]. In turn, these theoretical findings, which have withstood the test of time, led to many other theoretical and empirical advances in different fields of science. This book attempts to glean from several of these advances in the following chapters, yet the original study that inspired these theoretical advances was based on compromised data.

How can a good theory come from spurious data? Reviewing the literature, it becomes apparent that although wandering albatrosses may not in fact perform Lévy walks (although debate continues), it seems that many organisms diffuse anomalously (see Part II). The albatross data were corrupted, yet they led to new ideas that subsequently led to significant advances.

In this respect, the albatross study is not unique. Recall, for example, how Bohr's theory of the atom was wrong, yet it helped pave the way toward Schrödinger's wave mechanics and Heisenberg's matrix mechanics, unified into a general theoretical framework by Dirac. Similarly, corpuscular theories of light of the seventeenth century were later empirically shown to be inadequate, although in retrospect, they were a step toward a deeper understanding of optics. In such cases, scientific progress itself resembles a random walk. The albatross study is just one example of this.

We now know a lot more about how wandering albatrosses move. For example, Fritz *et al.* [125] have shown that wandering albatrosses use a consistent pattern of movement. Specifically, they apparently use three scale-dependent nested domains where they adjust tortuosity to different environmental and behavioral constraints. At small scales of about 100 m they use a zigzag movement as they continuously adjust for optimal use of wind. At a medium scale of 1–10 km, the movement shows changes in tortuosity consistent with food-searching behavior. Finally, at large scales greater than 10 km, the movement corresponds to commuting between patches and is possibly influenced by large-scale weather systems. Hence, although the Lévy flight idea is too simple to account for all the complexity seen in the real data, the Lévy flight idea nevertheless helped to draw attention to scale invariance and the importance of analyzing data at multiple scales.

Part II

Experimental findings

5

Early studies

5.1 Fickian transport

The classic paradigm of simple diffusion is used to describe a wide range of phenomena, ranging from how the original humans migrated and dispersed out of Africa to the spread of pollen. Until the twentieth century, Fick's laws were thought to be universally valid for describing diffusion. The physiologist Adolf Fick introduced the idea that diffusion is proportional to the gradient of concentration. For practical as well as for historical reasons, normal diffusion is commonly assumed for transport processes. For example, Fourier's law for heat flow is analogous to Fick's laws of diffusion, with temperature gradients playing the role of concentration gradients.

Like Gaussian statistics, normal diffusion is ubiquitous because of the wide applicability of the central limit theorem. Standard methods in spatial ecology traditionally have tended to assume Brownian motion and Fickian diffusion as two basic properties of animal movement in the long time limit, i.e., at large spatial scales and long temporal scales. We refer the reader to the seminal book by Berg [35] on random walks in biology.

Fickian or normal diffusion assumes that animal movements can be modeled, in the long-term limit, as uncorrelated random walks [21, 35, 265]. In many cases, normal diffusion describes experimentally observed phenomena. The classic study by Skellam [349] of the colonization of Europe by muskrats assumed normal diffusion, for example (Figure 5.1). Similarly, the motility of the predominantly surface-dwelling ciliated protozoon *Euplotes vannus* can be described as a classic two-dimensional random walk with a Poisson distribution of run lengths, punctuated by random changes in walking direction [117].

However, the difficulty with such uncorrelated random walks is that they do not account for directional persistence. Real organisms have a tendency to continue moving in the same direction. For example, animals (and people) only rarely

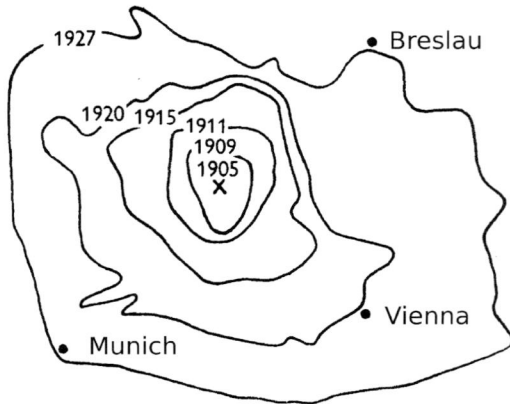

Figure 5.1   Skellam [349], in the early 1950s, studied how muskrats spread in central Europe. The figure shows how the colonized area increased with time. In the 1900s, muskrats were brought to Europe for fur breeding. Five muskrats were accidentally released in 1905, near Prague. The release location is marked by an X in the figure. Within half a century, they had colonized most of the continent. Skellam used normal diffusion together with exponential population growth to model the spread and predicted that the square root of the colonized area would grow linearly in time. The theoretical prediction agreed spectacularly with the empirical data, and today, the original paper [349] is considered a classic. (Adapted from the 1951 figure by Skellam [349].)

make 180 degree turns (or, equivalently, walk backward). However, uncorrelated random walks assume – often unrealistically – that it is as equally probable that an organism will make a 180 degree turn as a 1 degree turn. Such limitations have been addressed using two types of random walks: short-range correlated random walks and superdiffusive walks [21].

5.2 Directional persistence

In the 1980s and early 1990s, it became apparent [21, 42, 173] that uncorrelated random walks do not realistically describe actual animal motion. In an uncorrelated random walk, each random walk step is statistically independent. In contrast, if a real animal is moving in a given direction, the odds favor continuing in that direction. Why take two steps forward and one back when you can take three steps forward? Revisitation of previously visited sites, known as oversampling (Figure 5.2), wastes time and energy.

The simplest way to incorporate directional persistence into a random model is to introduce correlations (i.e., memory effects) between successive random walk steps. Correlated random walk (CRW) models are discussed in greater detail in Chapter 11 (see also Figure 11.1, illustrating a CRW). For now, we only mention

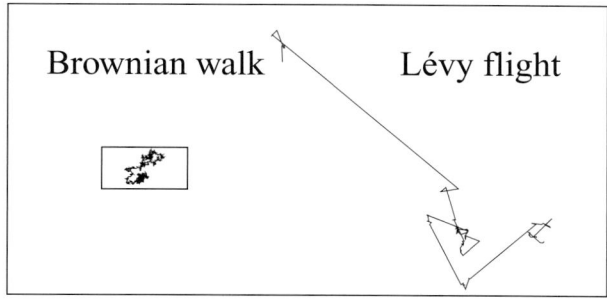

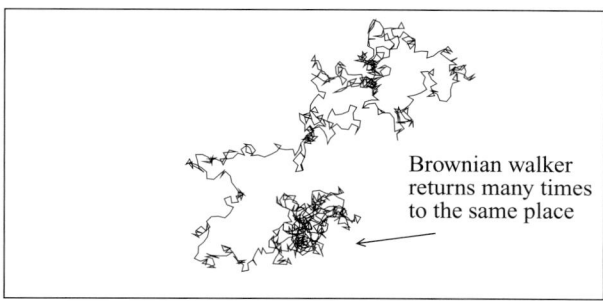

Figure 5.2 Two-dimensional Brownian random walk and Lévy flight of identical total length of 1000 units, shown to scale. (Bottom) Zoom of the Brownian walk. Note how the Brownian walker returns many times to previously visited locations, a phenomenon known as *oversampling*. In contrast, the Lévy flyer frequently takes ultralong jumps to virgin territory. This reduction in oversampling is part of the fundamental theoretical basis for interest in the Lévy flight foraging hypothesis.

that the trajectories generated by CRW models appear more similar to the empirical data than those generated by uncorrelated random walks.

CRWs thus represent an advance in the quantitative description of animal movement. However, the correlation function for CRWs decays exponentially; hence the memory effects have a finite range, and beyond certain spatial and temporal scales, CRWs become uncorrelated random walks. In other words, the correlations are not strong enough – there is a loss of directional persistence at large spatial and temporal scales.

5.3 A new idea: Lévy flights and walks

It was in this historical context that the idea of Lévy flight foraging was born. In contrast to CRWs, Lévy flights and superdiffusion can account for long-range dispersal (Figure 5.3). The earliest reference containing the idea that biological organisms could perform Lévy walks is a theoretical paper by Shlesinger and

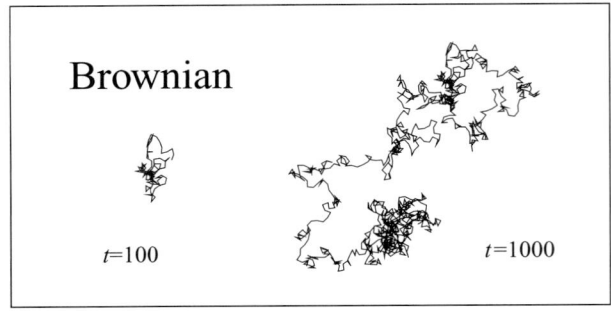

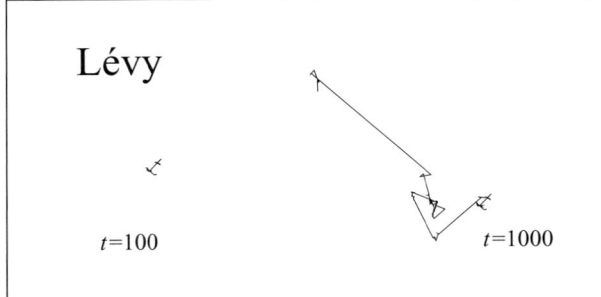

Figure 5.3 Increasing the length of a Brownian random walk by a factor of 10 only leads to an increase of the order of $\sqrt{10}$ in the linear size of the visited region. In contrast, a 10-fold larger Lévy walk can yield up to a 10-fold increase in the linear size due to the reduced oversampling. Hence it makes sense to consider Lévy walks and superdiffusion in the context of animal movement. Somewhat surprisingly, the idea first appeared in the literature only in the 1980s (see the text), about half a century after Paul Lévy discovered the stable distribution associated with his name.

Klafter [341], in 1986. The earliest experimental investigations are the plankton ecology papers by Levandowsky *et al.* [205, 206], in 1988, about the swimming behavior of micro-organisms.

Subsequent experimental studies of Lévy motion appeared in print circa 1995–1996 [85, 207, 332, 390]. One of the first studies of the very large-scale foraging motion of organisms, of the wandering albatross, is discussed in Chapter 4. Cole [85] analyzed the episodes of activity and rest of *Drosophila melanogaster* and found a fractal (i.e., self-similar or scale-invariant) structure, concluding that the fractality could lead to Lévy flight patterns of movement in fruit flies. Schuster and Levandowsky [332] used time-lapse videomicroscopy to characterize the chemotactic behavior of axenically grown *Acanthamoeba castellanii*. They concluded that the diffusive behavior did not support Brownian motion, raising the possibility of Lévy walks. A subsequent study, by Levandowsky *et al.* [207], confirmed the earlier findings of superdiffusion.

In 1999, motivated by such empirical findings, we proposed [393] what later became known as the Lévy flight foraging hypothesis, based on our finding that (truncated) Lévy flights optimize random searches under certain circumstances. The central idea of the hypothesis is that organisms must have evolved via natural selection to exploit this optimal strategy. Since then, a number of researchers have studied diverse organisms from a variety of perspectives. Indeed, the idea that some organisms may diffuse anomalously is no longer considered far-fetched or exotic. We catalog some of these results in the next chapter.

6

Evidence of anomalous diffusion

In addition to the early studies we discussed in Chapter 5, here we review other experimental studies. We focus on animal movement because of the relative abundance of data and larger velocities relative to plant seeds, pollen, etc. We note, however, that seed shadows are often fat tailed [408] (i.e., leptokurtic), so many plants may also disperse superdiffusively.

6.1 Arthropods and mollusks

Honey bees, fruit flies, and desert ants

Reynolds *et al.* [306, 307] studied bees using a variety of techniques. In one study, they [306] used harmonic radar to record the flight paths of honey bees that were searching for their hives. Harmonic radar can differentiate between the harmonic signal returned from a specific target and signals returned from all other passive reflectors. Analysis of the trajectories indicated scale-invariant walks with a power law exponent $\mu \approx 2$ (see also [300]), corresponding to a Lévy index $\alpha = 1$. They argued that these results, combined with the *no preferred direction* characteristic of the segments, demonstrate that the bees were flying an optimal search pattern. An inverse square power law distribution ($\mu = 2$) is exactly what the theory of Lévy flight foraging predicts (see Chapter 10).

In another study, Reynolds *et al.* [307] trained foraging honey bees to seek out an artificial feeder, which was subsequently removed. The resulting bee flight patterns were recorded using harmonic radar and showed that the flight patterns have the scale-free characteristic of Lévy walks. They argued that this pattern constitutes an optimal searching strategy for the location of the feeder and demonstrated that this searching strategy would remain close to optimal even for imperfect implementation of the walks. Reynolds [302] has also shown that $\mu = 2$ optimizes central place foraging via a power law distribution of loop lengths and has suggested that the searching patterns of desert ants (*Cataglyphis*) are consistent with an optimal Lévy-looping searching strategy [302].

Reynolds and Frye [304] studied trajectories of fruit flies in still air and found evidence suggesting optimal scale-free searching and an inverse square power law distribution of inter-saccade interval lengths. They also found evidence of intermittent searching (see Chapter 11) of alternating search and relocation phases. This study constitutes an example of searching behavior that is both scale free and intermittent.

Meats and Edgerton [236] studied the dispersal of immature and sexually mature Queensland fruit flies, *Bactrocera tryoni* (Froggatt), from recapture (i.e., trap) rates after releases made at a single point. They reported a scale-invariant distribution for long-distance dispersers, consistent with a power law tailed probability density function.

Butterflies and moths

Reynolds also speculated that Lévy searches might underlie visually cued mate location in butterflies [298]. In a separate study, Reynolds *et al.* [305] studied the behavior of *Agrotis segetum* moths and found complex flight patterns compatible with scale-free Lévy flight searches.

Root-feeding insects

Given the relative ease of movement above ground, and the theoretical plausibility of Lévy flight movement patterns, evidence of such behavior for above-ground organisms should not cause surprise. Intriguingly, however, the same may hold true for below-ground organisms. Johnson *et al.* [169] and Zhang *et al.* [417] studied the movement of the clover root weevil (*Sitona lepidus*) and reported that the results suggest Lévy walks. Given the methodological difficulties associated with studying these insects in the soil and the lack of well-developed empirical and theoretical frameworks, the systematic study of the movements of root-feeding insects shows promise for new and interesting results in the future.

Snails

Seuront *et al.* [337] studied the two-dimensional motion of gastropod *Littorina littorea*. They found that the probability density function of displacements has a power law tail, with exponents between $\mu \approx 2.2$ and $\mu \approx 2.7$, corresponding to the superdiffusive regime. Note that there is no contradiction between the slow speed of a snail, and a superdiffusive snail.

6.2 Marine and aquatic animals

Sharks

Lévy flight search patterns possess structure across multiple scales, as mentioned above. Sims *et al.* [345] analyzed the relative success of shark searches by comparing prey (zooplankton) biomass encountered by sharks in a dynamic prey landscape (in the northeast Atlantic Ocean) with encounters by random walk simulations of model sharks. They found results "consistent with basking sharks using search tactics structured across multiple scales."

Gray seals

Austin *et al.* [12] studied 52 gray seals fitted with satellite-linked recorders on Sable Island, located southeast of mainland Nova Scotia in the Atlantic. They found considerable variation in types of movement, and the majority of seals, in fact, do not adopt Lévy walks. Interestingly, however, a large minority of seals, up to 30% of the gray seals studied, have a Lévy distribution of move lengths. The reasons underlying such intraspecific variation in the type of movement are not yet well understood.

Stream fish

Zhang *et al.* [417] studied a fresh water (stream) fish (*Cyprinidae: Nocomis leptocephalus*) and found possible evidence of leptokurtic movement, i.e., movement described by fat-tailed probability density functions for displacements.

Bony fish, sharks, sea turtles, and penguins

The most convincing evidence of Lévy walks includes a statistically rigorous study of bony fish, sharks, sea turtles, and penguins by Sims *et al.* [347], published in 2008, and an even larger follow-up study [163] that reconfirmed the results in 2010 (see Section 14.5).

Sims *et al.* analyzed data sets comprising more than 10^6 data points and found strong evidence of Lévy processes. Their use of rank-frequency plots and comparisons of different models takes into account recent criticisms of the use of double-log plots of histograms, e.g., by Edwards *et al.* [105]. An important point that emerges from such studies is that not all animals use Lévy walks. Instead, Lévy walks appear to fall within the spectrum of observed behaviors. The second study, by Humphries *et al.* [163], used more than 10^7 data points.

Sims, Humphries, and their collaborators contributed significantly toward settling the debate about whether animals diffuse anomalously. Instead, the debate now is about which, when, and why some animals move in this manner (see Section 6.8).

6.3 Mammals

In addition to the aquatic mammals discussed earlier, the following other mammals have been studied and appear to adopt anomalous diffusion.

Reindeer

Mårell *et al.* [224] studied foraging-site selection by the semidomesticated female reindeer (*Rangifer tarandus tarandus L.*) and found discrepancies between the actual reindeer foraging paths and a (non-Lévy) correlated random walk model.

Deer

Early findings [393] of Lévy flight patterns of deer required correction [105], so the behavior of deer in this context once again became the topic of discussion. However, by using recently proposed statistical methods, Focardi *et al.* [121] reported that fallow deer (*Dama dama*) actually perform Lévy walks while searching alone for food, with exponent $\mu = 2.16 \pm 0.13$. This was perhaps the first account of Lévy foraging of a terrestrial mammal using up-to-date statistical analysis methods. Curiously, Lévy foraging behavior could not be observed in clustered deer.

Jackals

Atkinson *et al.* [11] used radio-tracking techniques employed by field ecologists to look for fractal behavior in the foraging trajectories of the African side-striped jackal. They not only reported evidence of superdiffusion, but also showed the complexity of actual foraging trajectories, which includes curvature (or equivalently, a degree of sinuosity).

Spider monkeys

Ramos-Fernández *et al.* [292] and Boyer *et al.* [46] studied the foraging patterns of free-ranging spider monkeys (*Ateles geoffroyi*) in the forests of the Yucatan Peninsula and found a power law tailed distribution of steps consistent with a Lévy walk. The study reported, moreover, a power law distribution for waiting times, which could indicate (though not explicitly mentioned in [292]) a Scher-Montroll type of continuous time random walk [250, 326, 327]. They also found strong evidence of superlinear scaling of the mean squared displacement ($H > 1/2$), i.e., evidence of superdiffusion. A subsequent study by Boyer *et al.* [47] interpreted these findings in the context of the interaction of the foragers among them and with complex heterogeneous environments.

Elephants

Dai *et al.* [94] studied a herd of elephants in South Africa and found evidence that they follow a recently proposed stochastic process known as the Lévy-modulated correlated random walk [21]. These differ from the standard correlated random walk in that Lévy-modulated correlated random walks are genuinely superdiffusive across multiple spatial and temporal scales, whereas correlated random walks appear superdiffusive (ballistic) only at short scales. The standard correlated random walk is diffusive at large scales.

Goats

A study by de Knegt *et al.* [97] of the movement strategies of goats is interesting for several reasons. The authors demonstrate that animal movement becomes slower and more tortuous when the patch density (of trees and bushes) increases. Specifically, the net displacements are smaller for larger patch densities, as expected from the theory of optimal foraging. Why travel far away when food is near? What is even more remarkable, however, is that the power law exponent is approximately $\mu = 2$ for low patch densities but approaches $\mu = 3$ for high patch densities. This relationship between density and the measured exponent μ is precisely what the theory of Lévy flight foraging predicts (see Chapter 10).

6.4 Micro-organisms

Dinoflagellates

An important study in the context of biological Lévy flight foraging was undertaken by Bartumeus *et al.* [20]. The study investigates the flight times of the dinoflagellate *Oxyrrhis marina*. They found that the distribution of flight times switched from an exponential (diffusively equivalent to $\mu > 3$, $H = 1/2$) to an inverse square power law distribution ($\mu = 2$, $H > 1/2$) when the prey (*Rhodomonas sp.*) decreased in abundance, as predicted theoretically (Chapter 10). In the late 1990s, the present authors had reported a similar dependence of the measured μ on the density of resources [393] (for bumble bees). Given this striking similarity between the studies of goats (see above), dinoflagellates, and bumble bees, this behavior of μ could be a more general phenomenon.

Bartumeus *et al.* [20] also studied the amplitude and frequency of the short-term helical path, which increased with decreasing density. They suggested that for a three-dimensional environment, a stronger helical component combines with Lévy search strategy to enhance the relevant encounter rates. They concluded that their results support the idea of a universally applicable statistical law of foraging.

6.5 Birds

A recent unpublished study (preprint, 2010) analyzes the large-scale movement of several species of birds and finds power law tailed distributions of displacements over extended time windows of up to one month.

Similarly, Bartumeus *et al.* [24] studied Mediterranean seabirds and found that when operating trawlers are present, their flight length distributions decay exponentially. They also reported that when trawlers are absent, the seabirds appear to move superdiffusively.

Penguins were discussed earlier, in the context of marine animals.

7

Human dispersal

This chapter reviews the evidence in favor of anomalous diffusion in the movement of human beings. Human diffusion constitutes a driving force for various spatiotemporal phenomena that occur on large geographical scales [52], and can synchronize and stabilize populations as well as diversify gene pools.

The spread of infectious disease, for example, depends on human diffusion. Consider, for example, the spread of swine flu (H1N1 influenza A [56]), resulting in the World Health Organization raising the pandemic alert to phase 6 – the highest level – in June 2009, only a few months after the virus's first appearance in Mexico. It is difficult to model or understand the pandemic in terms of normal diffusion or the kinds of wave fronts seen in solutions of the standard (i.e., nonfractional) Fisher-Kolmogorov equation. The virus appears to jump across continents and oceans in a very short time, in a manner more commensurate with superdiffusion than with normal diffusion. Similarly, the 1918 flu pandemic (Spanish flu) lasted only a couple years but reached nearly all corners of the planet with a rapidity consistent with a superdiffusive process mediated by human mobility.

Given the role of human diffusion – or superdiffusion – in pandemics and other reaction-diffusion processes, we now examine how humans diffuse. Because we assume that air, sea, and highway transportation increase mobility and diffusivity, we begin by looking at human societies that predate those modern modes of travel.

7.1 Hunter-gatherers and archaeological evidence

The study by Brown *et al.* [58] of the distribution of distances between campsites of Dobe Ju/'hoansi hunter-gatherers provides a conceptual advance. The foraging patterns of the Dobe Ju/'hoansi may in fact follow a Lévy walk (or some other form of anomalous diffusion).

Until very recently, the Dobe Ju/'hoansi were hunters and foragers living in and around the Kalahari Desert, in present-day Botswana and Namibia. This region partially overlaps the habitat of desert elephants [204] (see Section 6.3 for a discussion of African elephants [371] and their long-distance movements [370], including Lévy walks). The Dobe Ju/'hoansi are perhaps best known for their click language of the Khoisan family (Ju/'hoansi is pronounced approximately as "zhutwasi" or "juntwasi" in English). It is commonly believed that Khoisan peoples inhabited much of southern Africa before the Bantu expansion from the north. Their skeletal remains bear similarities to those who lived there as far back as the late Stone Age. Hence the finding that the Dobe Ju/'hoansi diffused anomalously suggests that their ancestors – and perhaps ours as well – dispersed superdiffusively.

We also note that Lévy walks have been used to model forager mobility based on archaeological sites [51]. Similarly, Brown *et al.* [57] explored fractal concepts in archaeological site patterns.

7.2 Lévy flights of dollar bills

In an innovative study of anomalous diffusion and transport due to human travel, Brockmann *et al.* [56] studied bank note circulation, using it as a proxy for the movement of people. They found two distinct deviations from normal diffusion in the transport of humans. First, they found power law tails in the probability density function of waiting times in small regions. Taken in isolation, such power law distributed pausing times in CTRWs lead to subdiffusion. They also found a power law distribution of travel distances, consistent with a Lévy flight pattern of movement. The effective power law exponent $\mu = 1.59$ lies in the superdiffusive regime and corresponds to a Lévy index of $\alpha = 0.59$.

The opposing superdiffusive and subdiffusive tendencies compete for dominance, and the end result seems to be an attenuated superdiffusion. Indeed, Brockmann [52] showed that the empirical data for bank note diffusion agree with a superdiffusive two-parameter CTRW model with a surprising accuracy. Because dollar bills move only when carried by people, it suggests that the movement of people is also superdiffusive.

7.3 GPS tracking of humans

González *et al.* [137] studied the trajectories of 10^5 mobile phone users and found that human trajectories show a high degree of spatial and temporal regularity. Underlying all the variation and diversity, humans appear to follow a simple pattern. Specifically, González *et al.* found that each individual has a (nonunique) time-independent characteristic travel distance as well as a significant probability

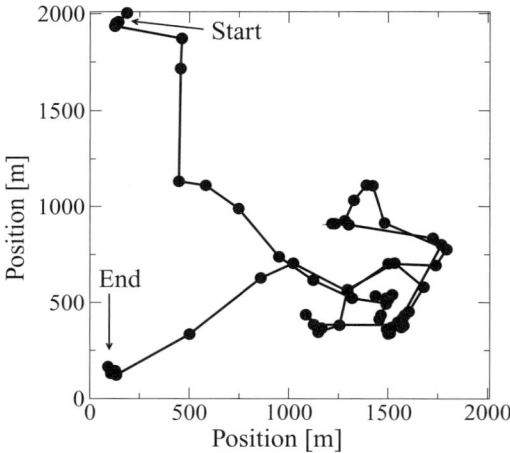

Figure 7.1 Rhee *et al.* [310] have studied, using GPS technology, how humans move. This plot shows an example from their study of human movement patterns. Compare the behavior shown here with the behavior of spider monkeys (Figure 1.1). The movement patterns of humans may share qualitative similarities with those of nonhuman primates.

of returning to a few highly frequented locations. After compensating for the variability among individuals, the data collapse onto an exponentially truncated Lévy flight [137]. At scales much smaller than the cutoff, their observed power law exponent $\mu \approx 1.75$ corresponds to a Lévy index $\alpha_L = 0.75$, consistent with superdiffusion.

Similarly, Rhee *et al.* [310] analyzed approximately 10^3 hours of GPS traces involving 44 volunteers. They found a truncated power law distribution consistent with a (truncated) Lévy walk. Notice that the human trajectories shown look surprisingly similar to those of spider monkeys – compare Figure 7.1 with Figure 1.1. It has become clear in the last few decades that when it comes to biological issues – including how we move – humans do not fundamentally differ from other animals.

7.4 Fishermen as foragers

In an intriguing experimental study, Bertrand *et al.* [39] found that the trajectories of Peruvian purse seiners follow a Lévy walk. The possibility that fishing boats might adopt a Lévy walk search strategy analogous to top predators supports the idea that people (unconsciously) adopt superdiffusive search strategies. In a subsequent study, Bertrand *et al.* [40] confirmed their earlier finding and noted that technology and communication do not seem to fundamentally affect this basic Lévy walk strategy (note that the power law exponent is close to $\mu = 2$).

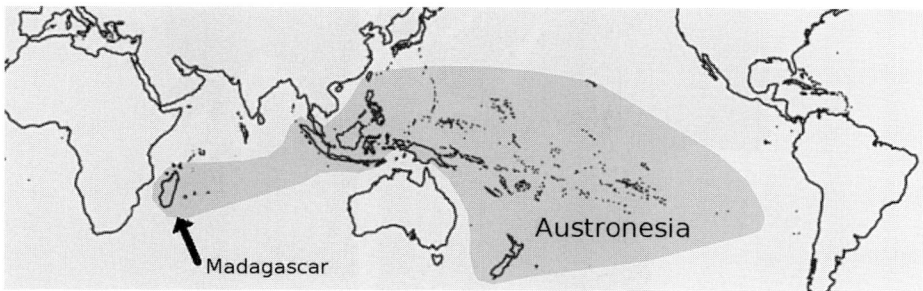

Figure 7.2 Approximate extent of Austronesia, the homeland of people who speak Austronesian languages [93, 99, 134, 140]. It is thought that they diffused out of prehistoric Taiwan, eastward to Easter Island and westward to Madagascar. Seafarers arrived in Madagascar approximately 15 centuries ago, possibly from present-day Indonesia. The incredible journey of more than 5000 km was undertaken on outrigger canoes [101].

Marchal *et al.* [223] studied the foraging of fishermen from the perspective of Lévy flight theory by examining case studies of North Sea Dutch and French vessels. Both fleets appeared to move in a manner similar to Lévy walks. They found that the Dutch vessels foraged in a manner consistent with $\mu \approx 1.5$.

7.5 Austronesians in Madagascar

The studies listed above make a strong case for the superdiffusive movement of humans, not only today but also in the remote past. Apart from the question of its validity, this hypothesis has an intriguing significance. It is commonly assumed that humans colonized the planet via normal diffusion. What if, instead, this colonization was superdiffusive?

A striking example of superdiffusion is the colonization of Madagascar by Austronesians circa A.D. 500 [93, 99, 134, 140] (Figure 7.2). The Austronesian languages diffused from Taiwan eastward toward Easter Island and westward toward Madagascar [99, 134, 140] – an incredible span of more than 25 000 km, approximately half the planet's circumference. These able seafarers from southeast Asia traveled more than 5000 km across the Indian Ocean, possibly in outrigger sailing canoes [101].

Assuming either normal (i.e., Brownian) diffusion or reaction-diffusion, the original Austronesian expansion that reached the island of Madagascar should have then promptly reached continental Africa, but there is little or no evidence that it extended to the east coast of Africa (Figures 7.2 and 7.3). Moreover, if the expansion process were governed by normal diffusion, the nearby islands would have been colonized rapidly (Figure 7.4), soon after – or even before – the initial

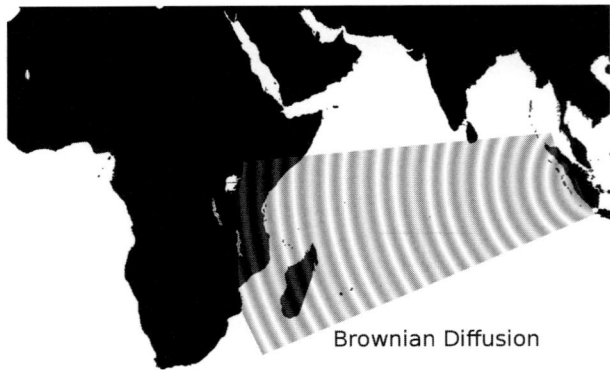
Brownian Diffusion

Figure 7.3 If the Austronesian immigration to Madagascar was due to normal (Brownian) diffusion, then the probability of reaching the east coast of Africa would not be much lower than the probability of reaching Madagascar, starting from Indonesia. There is little or no evidence that Austronesians reached the east coast of Africa during the original immigration. Moreover, normal diffusion is not characterized by rare but extreme events.

colonization of Madagascar. Assuming normal diffusion, it should have taken 100 times less time to diffuse 500 km than the time taken to diffuse 5000 km. Such a rapid colonization of the nearby islands did not take place, however. Indeed, several centuries were to pass before the Austronesians colonized the nearby islands (Figure 7.4).

Consider the following: (1) the Seychelle Islands were colonized by Arabs (in the seventh or eighth century), and then many centuries later by Europeans, but apparently not by the original Austronesians or their early local descendants; (2) Mauritius and (3) Reunion Island were not colonized until the tenth century, half a millennium after Madagascar. These and other similar timing discrepancies cannot be easily explained by the hypothesis that the colonization of Madagascar occurred via normal diffusion. What is even more important, atypically large and ultralong displacements do not occur in normal diffusion.

Brownian random walkers do not take giant steps, but Lévy walkers do. The hypothesis that the Austronesians reached Madagascar via anomalous diffusion can adequately account for the relevant facts (Figure 7.5). Assuming a Lévy walk, for example, rare but ultralong jumps are possible. Trade winds and the south equatorial current could have facilitated – or perhaps even induced – the ultralong 5000-km journey to Madagascar. Anomalous diffusion can also account for the relative slowness of colonization of the nearby islands. For a Lévy walk with $\mu = 2$, for example, it takes 10 times less time to diffuse 500 km compared to the time taken to diffuse 5000 km – not 100 hundred times less time, as happens

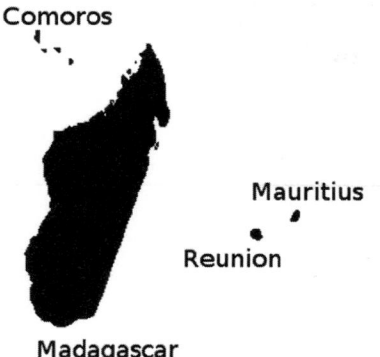

Figure 7.4 Soon after their arrival at Madagascar, Austronesians did not reach the many nearby islands; rather they reached these other islands only much later, if at all (see the text). If they reached Madagascar via normal diffusion from as far away as Indonesia, they should have reached the nearby islands very soon after – or even before – their arrival at Madagascar. Instead, could the Austronesians have reached Madagascar in one of the rare but extreme events typical of Lévy flights and walks?

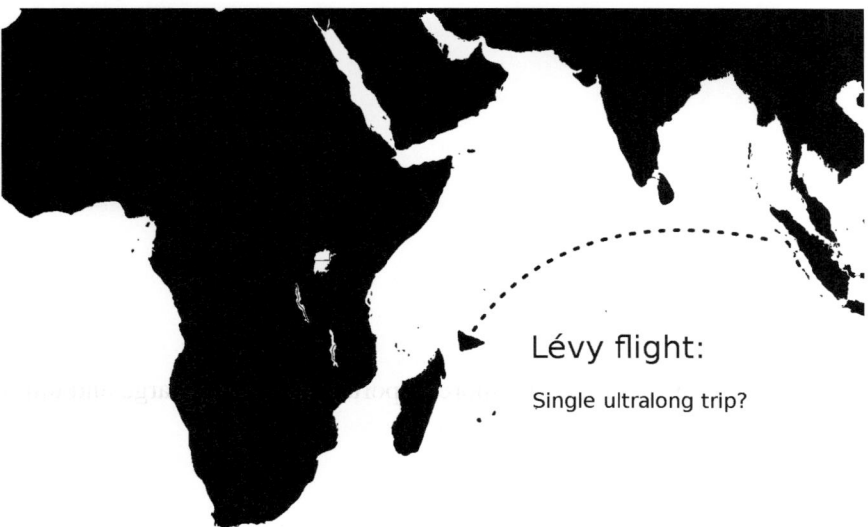

Figure 7.5 Lévy flights and walks are characterized by rare but extreme, ultra-long jumps, so the Lévy flight hypothesis can account for the 5000-km journey from Indonesia to Madagascar. Moreover, it also can account for the absence of Austronesian colonization of the east coast of Africa: the journey to Madagascar would have been a single event, albeit an extreme one. As explained in the text, there is growing evidence that humans diffuse anomalously.

in normal diffusion. Such effects may also account for the lack of evidence that Austronesians promptly reached and settled on the east coast of continental Africa. Perhaps a rare, ultralong Lévy flight was truncated on finding Madagascar.

It may seem counterintuitive that superdiffusion can leave large gaps or holes that only slowly get filled with time, but such effects are real. Unlike normal diffusion, superdiffusion tends to favor larger displacements and inhibit shorter displacements. Superdiffusion thus moves the walker far away from the starting point but leaves large, unexplored gaps. The physical explanation of this effect is that the fractal dimension of a superdiffusive random walk is lower than for a Brownian random walk. Lower fractal dimensions mean that the embedding space will be less filled by visited points, producing more gaps.

In summary, superdiffusion can account for both the extraordinary distance between Madagascar and southeast Asia and for the lack of rapid colonization of nearby islands and continental Africa. It is possible that the arrival in Madagascar of Austronesians is an example of the rare but ultralong jumps characteristic of Lévy flights and walks.

8

How strong is the evidence?

Studies of Lévy flight search patterns have come under strong scrutiny, especially after the events discussed in Chapter 4. However, the number of reports of Lévy flights and walks in animal movement continues to grow, as discussed in Chapter 6. Here we briefly review the case for and against Lévy flights (and, more generally, superdiffusion).

8.1 Measurement and data analysis

Experimental science is always susceptible to measurement problems. Bradshaw *et al.* [50] systematically studied the effect of ignoring errors when measuring animal trajectories and found that such errors may lead to substantial biases when interpreting movement patterns. As Kölzsch and Blasius [188] point out, the effects of seasonal drift on the migratory patterns of birds render random walk methods useful only if they are modified and interpreted carefully to take seasonal drift into consideration. Nevertheless, the following facts support the case for anomalous diffusion: (1) a large number of studies (listed earlier) have all reached the same conclusion and (2) the authors have used a number of measurement techniques and a variety of experimental setups. These facts, taken together, convey a degree of confidence in the quality of the data and the reliability of the reported conclusions.

A more difficult question concerns not data quality but how the data should be analyzed. Power laws in physical systems (e.g., earthquakes) typically scale over at least 3 orders of magnitude. For biological systems in general, and foraging dynamics in particular, even 2 orders of magnitude of scaling can be a luxury. Over a sufficiently small scaling range, any smooth function will appear as a straight line on a double-log plot. Sims *et al.* [346] examined ways of minimizing errors when identifying Lévy flight search patterns of organisms and compared different graphical methods of estimating the power law exponent μ. They found that cumulative histograms and logarithmically binned histograms provide better estimates than simple histograms. This is not a trivial problem and remains the subject of ongoing investigation. We are coauthors of a study, by Edwards *et al.* [105], that raises some concerns about using straight lines on double-log plots to infer Lévy walks (and power laws in general). Nevertheless, statistical inference methods also become problematic if truncation of the power law tails is not explicitly taken into consideration, a topic we discuss in Appendix A. Indeed, Nature abhors bare singularities and diverging moments, as opposed to *dressed* singularities or truncated power laws.

8.2 Special issues related to power laws

As an illustrative example, consider the studies of wandering albatrosses discussed in Chapter 4. Subsequent independent studies by Boyer *et al.* [48] and Reynolds [299] argued that the measured flight distributions for albatrosses do not in fact disagree with the hypothesis of a truncated Lévy walk. Indeed, it is well known that truncation effects are generated by finding targets [393, 397]. Specifically, the mean flight length $\langle \ell \rangle$ depends not only on the pure or *bare* power law tailed distribution $P(\ell)$ but also on the *mean free path* λ, which depends on the target density. This calculation, originally reported in 1999 [393], already shows that truncation of flight ℓ_j to length ℓ'_j guarantees that $\ell'_j \gg \lambda$ will be astronomically (i.e., exponentially) improbable [393, 397]. Indeed, the empirically measurable distribution of the flight lengths will not in general be identical to the probability density function of an underlying Lévy process, unless the target density is zero (i.e., no truncations).

Underlying distributions differ from the empirically measurable distributions due to a broken scale-invariance symmetry. Specifically, the perfect scale-invariance symmetry associated with the original bare power law tailed distribution becomes broken in the empirically observable distribution of the actual flights, because ultralong flights are truncated on finding targets. Thus, a characteristic scale emerges from the interaction with the environment, washing out or breaking the scale-invariance symmetry intrinsic to the underlying Lévy walk process.

8.3 Anomalous diffusion: Not if, but when and why

Benhamou [28] argues that even with unambiguous evidence supporting Lévy walks, this does not automatically imply an underlying Lévy walk process – the Lévy behavior "may emerge from the way the animal interacted with the environment structure through more classical movement processes" [28]. Instead, composite Brownian walks are proposed. In other words, Lévy walks and anomalous diffusion are side effects and not fundamental. In Chapter 13, we discuss this possibility in detail. Here we focus on whether any Lévy process – or any anomalous diffusion – actually occurs at all.

Reynolds [296] argued that *adaptive* Lévy walks with a power law exponent μ fluctuating between low values $\mu < 3$ between patches and large values $\mu \geq 3$ inside patches will appear indistinguishable from the composite Brownian walks of Benhamou [28], the latter having no underlying Lévy process.[1] In response to the comment of Reynolds [296], Benhamou [29] argues that Lévy processes with $\mu < 3$ necessarily possess a diverging (i.e., infinite) variance for distances. Because the trajectories of real animals will always include the effect of flights truncated due to found targets, the real trajectories will not represent true Lévy walks. In short, diverging moments are never observed in nature, and therefore the Lévy walk hypothesis is wrong.

Benhamou's argument against the equivalence of Reynolds's adaptive Lévy walks and correlated random walks is meaningless when applied to truncated Lévy walks instead of strict Lévy walks. We have already clarified, in Chapter 3, that nontruncated power laws are a physical impossibility and thus cannot exist in Nature. Carefully distinguishing between the underlying process and the empirically observable quantity is essential. Consider, as an illustrative example, charge singularities, e.g., in elementary particles. Even if they exist, the actual singularities are not observable. Similarly, the Gutenberg-Richter power law for earthquakes is truncated both above (due to the finite size of tectonic plates) and below (due to the nonzero size of real particles). For similar reasons, we cannot directly observe the pure underlying Lévy process. Nevertheless, the empirical observation of a truncated Lévy walk can imply an underlying Lévy process. Given the abundance of empirical evidence for truncated Lévy walks in animal locomotion, we cannot rule out this possibility. There is now a broad and growing acceptance of the more general view that organisms diffuse anomalously.

In summary, there is an abundance of evidence for anomalous diffusion generally and for superdiffusion ($H > 1/2$) specifically. Whether this superdiffusion arises from Lévy walks or some other mechanism is still an open question. The authors of this book strongly suspect that many biological organisms do in fact perform (truncated) Lévy walks. Good examples are the experimental studies by Reynolds *et al.* [306, 307] and Bartumeus *et al.* [20], which appear to provide

---

[1] Reynolds labels Lévy walks with time-dependent $\mu(t)$ as "adaptive Lévy walks." The word *adaptive* in this context points to the variation in μ and not to evolutionary adaptation.

robust results pointing in this direction. The most convincing studies are by Sims *et al.* [347] and by Humphries *et al.* [163], who analyzed more than one million movement displacements recorded from animal-attached electronic tags. The impressive amount of data shows that diverse marine predators perform anomalous diffusion. To paraphrase Bartumeus [373, 397], these works may well come to be seen as the studies that shifted the debate from "whether animals perform Lévy walks to when they use this strategy and why."

Part III
Theory of foraging

9

Optimizing encounter rates

The central idea underlying theoretical studies of the movement of organisms is that they need to *encounter* their targets. The targets can be other organisms of the same species (e.g., mates) or of a different species (e.g., prey) or, more generally, anything else sought (e.g., nesting sites). In the context of reaction-diffusion processes, the *reactions* (e.g., eating and mating) only take place when the relevant organisms successfully *diffuse* toward each other and meet. We next discuss a general theoretical approach to the study of encounter rates.

9.1 A general theory of searchers and targets

We classify the two interacting reactive-diffusive species (i.e., organisms) as either *searcher* (e.g., predator, forager, parasite, pollinator, male) or *target* (e.g., prey, food, female). Both searchers and targets move stochastically. We can now include most of the interactions in real ecosystems in this general framework [19], including the classical predator-prey interactions where an organism eats (usually smaller) organisms. It also includes diverse other interactions, such as osmotrophs looking for substrates and nutrients; parasites (including viruses) infecting organisms much larger than themselves (classical host-parasite interactions); organisms looking for aggregates (mixtures of amorphous organic matter, micro-organisms and/or inorganic particles), swarms, wakes, etc., also larger than themselves; and even mating encounters in which both male and female may have similar sizes (although sexual dimorphism is common) [19].

According to the theory of optimal foraging [128, 364], evolution through natural selection has led over time to highly efficient – even optimal – strategies. Therefore animals seek to maximize the returns (in calories, nutrients, etc.) on their labor when making foraging choices. In particular, organisms may move in a manner that optimizes or nearly optimizes encounter rates, and there is some experimental evidence that supports this possibility [329]. Because environmental and biological situations are highly variable, different optimized foraging strategies should naturally have evolved. We note, however, that there are alternatives to the optimal foraging theory, such as the regularity hypothesis [260], according to which animals prefer to stay inside a patch unless it becomes unprofitable. Laziness rather than optimization drives the foraging in this case.

The encounter rates between organisms are a limiting factor constraining activities such as feeding or reproduction. They are thus subject to evolutionary pressure and play a crucial role in the generation of biodiversity at evolutionary timescales. In a broad sense, an optimal foraging strategy should maximize encounter rates and minimize energy costs. Visser [387] has shown, in the context of the relative fitness of zooplanktonic organisms foraging under the risk of predation, that a convoluted swimming path with meanders, zigzags, and spirals confers greater fitness than rectilinear motion. The ability of organisms to change how they move relative to changes in the local environment can lead to unexpected behavior. For example, large herbivores increase their speed when crossing denser shrubby patches [10]. Contrary to conventional wisdom, the rate at which a predator encounters prey does not necessarily increase linearly with target density [374]. Pinaud and Weimerskirch [285] studied albatrosses and other marine seabirds and demonstrated that such predators adjust their foraging behavior according to what is happening in their heterogeneous environments. Such flexible foraging even occurs among apparently "simple" organisms moving in highly heterogeneous landscapes, e.g., symbiotic dinoflagellates in soft coral. Algae can effectively locate their hosts in both still water and flowing water, and their chemoreceptive abilities allow them to modify their swimming patterns, including direction, velocity, and turning rates [273]. The "convoluted" trajectories mentioned earlier are ideally suited to an analysis using random walks and fractal curves [336]. Uttieri *et al.* [379] studied the relationship between the encounters and the fractal dimensions of 3-D trajectories.

The successful theoretical modeling of random searches and other encounter processes relies on the fact that the rate of biological encounters is determined by the statistical properties of the movement (e.g., the type of diffusion) when all other things remain constant, such as chemical gradients, target density, learning, and information availability.

We next describe an idealized model that captures the most important statistical properties of the diffusion and preserves the main dynamical aspects of searches, in a limited context in which predator-prey relationships can be ignored and learning minimized [393, 397]. We put aside more complex landscapes in order to focus on how encounter rates depend on transport and diffusion properties. Complex

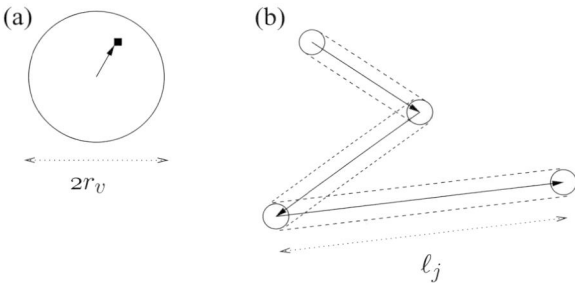

Figure 9.1 A limiting but general model of random searches [393]. (a) If there is a target site (solid square) located within a distance r_v, then the forager moves in a straight line to it. (b) If there is no target site within r_v, then the forager chooses a random direction and a distance ℓ_j from $p(\ell_j)$ and then proceeds as explained in the text.

landscapes and learning are also relevant [21, 47, 390], but lie beyond the scope of this book.

9.2 A limiting but general model of optimal foraging

An important goal in the physics of foraging is to capture the essential dynamics of encounter processes. We understand this in analogy to how the statistical physics of the Ising model of spin-$\frac{1}{2}$ particles captures the main features of ferromagnetism and critical phenomena. In 1999, we proposed the following general model [393], illustrated in Figure 9.1:

(1) If there is a target site located within a *direct-vision* distance r_v, then the forager moves in a straight line to the nearest target site.
(2) If there is no target site within a distance r_v, then the forager chooses a direction at random and a distance ℓ_j from a probability density function $p(\ell_j)$. It then incrementally moves to the new point, constantly looking for a target within a radius r_v along the way. If it does not detect a target, it stops after traversing the distance ℓ_j and chooses a new direction and a new distance ℓ_{j+1}; otherwise, it proceeds to the target, as in step 1.

Although the type of diffusion is determined by $p(\ell)$, the two rules above do not fully formulate the problem. Animal search models must also take into consideration the different kinds of targets. Do targets disappear or regenerate, i.e., is the foraging destructive or nondestructive? A nondestructive forager can revisit the same target site (many times). *Nondestructive searches* can occur in either of two cases: (1) the forager becomes satiated and leaves the site or (2) the target site is only temporarily depleted. In a *destructive search*, the target site found by the forager cannot be detected in subsequent searches. An intermediate case is *regenerative foraging*, in which the target is found, utilized, disappears, but then reappears and becomes available again after some large but finite characteristic time. Destructive searches correspond to the limit during which this regeneration time goes to infinity, whereas nondestructive search corresponds to the limit during which the regeneration time goes to zero.

Indeed, regenerative searches may be the most realistic type of search. For example, a tree that has been depleted of ripe fruit may regenerate as new fruit ripens over a period of time, which may be as short as a few days but not as short as a few minutes. Although the target regenerates and can be revisited, there is a significant refractory time during which it is not targetable.

9.3 Random walk propagators and encounter rates

In the next chapter, we will treat the above model quantitatively. Before we discuss the model in detail, we will examine the underlying physics. The quantitative results presented in the next chapter can be understood qualitatively and intuitively by considering the properties of the random walk propagator to be a function of the jump size probability density function $p(\ell)$. For walkers starting at the origin initially, random walk propagators (Figure 9.2) give the probability density function for random walkers at subsequent times. For noninteracting random walkers, and when the initial conditions are arbitrary, the principle of linear superposition can be used to express the probability density function for all future times in terms of propagators.

We first consider destructive searches in one dimension (1-D). Let time $t = 0$ and position $x = 0$ denote the time and position of the searcher immediately after the searcher finds a target. Assuming randomly distributed targets, the closest targets to the left and right of the searcher will be, on average, equally distant. This means that it does not matter whether the searcher moves to the left or right. Instead, all that matters for minimizing the total traveled distance is that once a choice is made to go left or right, this direction be maintained without change until the next target is found. There is no stochasticity, except possibly at $t = 0$. This type of movement between targets corresponds to ballistic motion and is not random. The propagator (Figure 9.3) consists of two Dirac δ-functions that move with equal velocity in opposite directions, starting from the origin $x = 0$ at $t = 0$ (Equation (3.41)). It is intuitively clear why ballistic behavior is best for destructive searches. In 2-D or 3-D destructive searches, the same argument holds, although the ballistic propagator is now a ring or sphere described by Dirac δ-functions

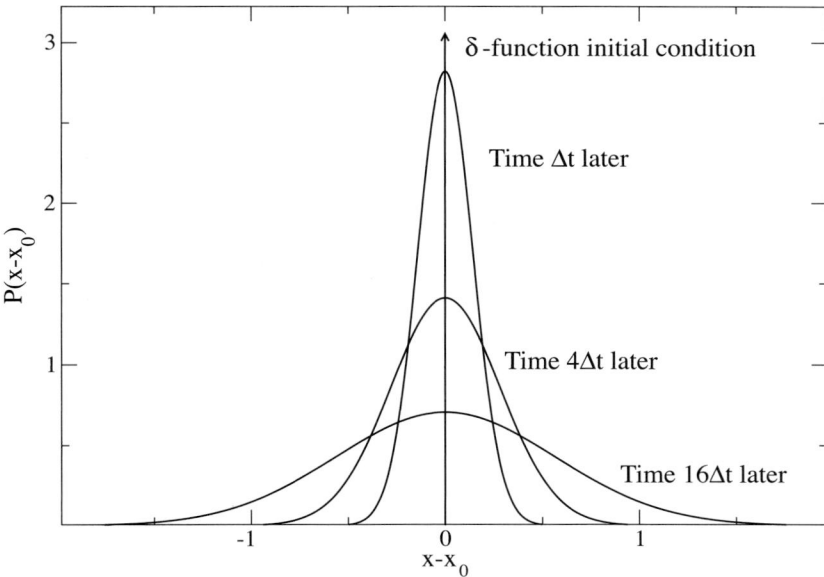

Figure 9.2 Random walk propagators $P(x, x_0; t, t_0)$ give the probability density to reach the position x at time t, starting at position x_0 at time t_0. For $t = t_0$, the propagator is a Dirac δ-function. Normal diffusion is described by Gaussian propagators $P(x, x_0; t, t_0) = P(x - x_0, t - t_0)$, whose variance grows linearly in time. The plot shows normalized Gaussian propagators at successive times, where Δt is an arbitrary time interval chosen to illustrate the evolution of the propagator.

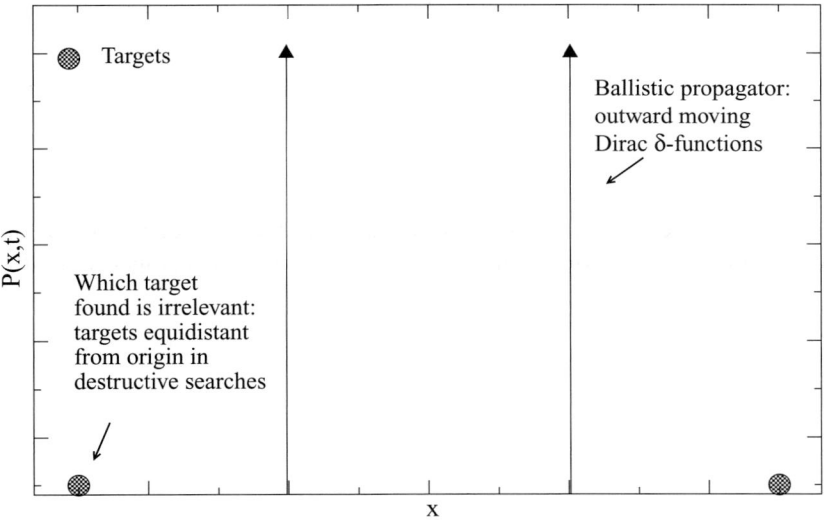

Figure 9.3 The ballistic propagator optimizes destructive 1-D searches because the targets to the left and right of the searcher are, on average, equidistant from the initial position of the searcher. Any time spent going backward is a waste.

in the radial variable. However, the main idea is not changed: it is best to keep moving away from the origin because there is no benefit in returning.

We next consider nondestructive searches in 1-D, where the target regenerates after some (very) short time. These differ from destructive searches in that at time $t = 0$, we can no longer assume that the searcher will be equally distant from the closest targets to the left and right. The previously visited target will regenerate quickly so that either the left or the right target will be closer, with the other target much farther away. This asymmetry will be more pronounced if the target density is low, and there may be orders of magnitude difference in the distances to the two targets. In this scenario, what is the optimal propagator for minimizing the total distance traveled to find the next target?

A Gaussian propagator for Brownian motion will bias the searcher toward the closer target because it will come within the range of the bell-shaped region of the propagator before the more distant target. In contrast, a ballistic propagator will not favor either target; the probability of finding the closer target is exactly 0.5 in 1-D. Both propagators cause inefficient searches, but for different reasons. The ballistic propagator moves the searcher to the next target quickly (in time proportional to the distance) but will often pick the farther target. By always moving to the closer target, the distance can be further minimized. Although the Gaussian propagator of normal diffusion preferentially takes the searcher to the closer target, the Hurst exponent is $H = 1/2$; hence the total distance traveled scales with the square of the distance to the target. For this reason, the Gaussian propagator is not very efficient. Specifically, random walks with $H = 1/2$ often return to previously visited sites before reaching an unvisited site. This *oversampling* [81, 213] of previously visited sites is the main drawback of normal diffusion and results in inefficient searches. If you have already checked that you did not leave your keys on the table, why return to the table to check again (Figure 5.2)?

Combining the superdiffusive property of ballistic motion with the ability to pick the closer target gives us the best of both worlds, a compromise represented by the Lévy walk. Lévy walk propagators contain Dirac δ-function spikes typical of ballistic propagators at the leading edge $x = \pm vt$, but more interesting is the presence of power law tails. The Dirac δ-function spikes can, in fact, be thought of as representing the portion of the power law tail that has been truncated due to effects of the finite velocity. To simplify the discussion, we illustrate the key ideas using the much simpler Lévy flight propagator. Although, at any given time, the power law tail in Lévy flight propagators extends to both the nearer and the more distant targets, the probability of reaching the closer target is much larger – for the same reason as for Gaussian propagators. Specifically, the fact that the propagator is larger near the origin and smaller farther out means that the closer target will

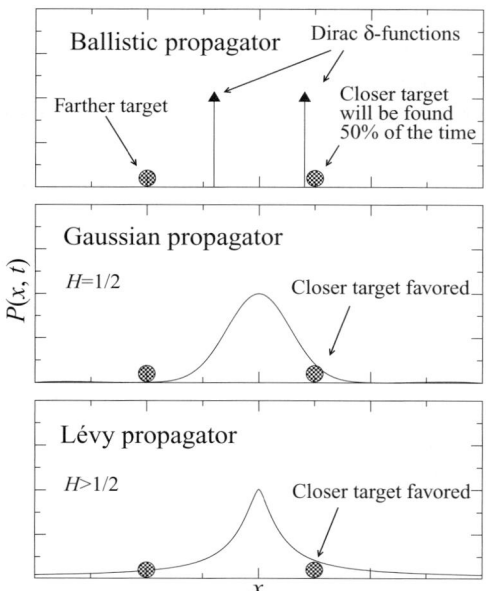

Figure 9.4 Illustrative sketches of 1-D random walk propagators for ballistic motion, normal diffusion ($\alpha = 2$), and superdiffusive motion reminiscent of Lévy flights ($\alpha < 2$) for the case of nondestructive foraging. The ballistic propagator, corresponding to the limit $\alpha \rightarrow 1$ of Lévy walks, consists of two Dirac δ-function spikes of area $1/2$ traveling in opposite directions (Equation (3.41)), and hence both left and right targets are equally likely to be reached by the searcher. Hence if one target is closer to the origin, the ballistic searcher wastes time and energy unnecessarily for 50% of the targets. In contrast, the diffusive Gaussian propagator is more likely to reach the closer target than the more distant target. In compensation, however, the Hurst exponent $H = 1/2$ is too low, so the search takes a time proportional to the square of the distance to the target, leading to low search efficiencies. Superdiffusive Lévy walk propagators, however, are a best-of-both-worlds compromise. On one hand, the Lévy propagator favors the closer target. On the other hand, large Hurst exponents $H > 1/2$ lead to less time taken to reach the target, relative to diffusive "Brownian" searches. Note that the leading edges of the Lévy walk propagator have slowly decaying Dirac δ-function spikes (not shown here to avoid confusion).

have a larger probability of being reached first. In contrast, the ballistic propagator is zero everywhere, except at the leading edge, where it is infinite.

We now see why Lévy flights represent a good compromise between ballistic and diffusive motion. On one hand, the Lévy flight and Lévy walk propagators preferentially lead the searcher to the closer target. This biasing effect is larger for the larger Lévy index α, with $\alpha = 2$ (the Gaussian case) being optimal for reaching the closer target. On the other hand, the movement time need not be proportional to the square of the distance to the target ($H = 1/2$). The largest possible Hurst exponent value for Lévy walks is $H = 1$, for $\alpha \leq 1$.

These considerations suggest why a Lévy index $\alpha = 1$ is the best compromise. The value $\alpha = 1$ allows the largest possible Hurst exponent for Lévy walks and simultaneously allows the searcher to reach the closer target preferentially. As the value of α decreases below $\alpha = 1$, the probability that the closer target will be chosen decreases and approaches 0.5 as $\alpha \to 0$. The same qualitative argument holds in high dimensions, except that instead of two directions (namely, left and right), there are now an infinite number of directions from which to choose. In the next chapter, we examine these qualitative arguments in quantitative terms.

10

Lévy flight foraging

10.1 The Lévy flight foraging hypothesis

The Lévy flight foraging hypothesis [393, 397] states that because truncated Lévy flights optimize random searches, biological organisms must have evolved to exploit Lévy flights. One central concept uses the power law exponent $\mu = \alpha + 1$ for the jump size distribution:

$$p(\ell) \sim \ell^{-\mu}, \quad \ell > \ell_0. \tag{10.1}$$

Variation of the parameter μ allows superdiffusive Lévy searches as well as Brownian searches involving normal diffusion. Here α is the Lévy index of the Lévy flight propagator, for the range $1 < \mu \leq 3$, and ℓ_0 is a lower cutoff where the power law tail begins. By comparing the search efficiencies as one varies μ, we can determine how much advantage can be gained by exploiting diffusivity and stochasticity in this way.

Note that because Lévy walks and flights – like Brownian motion – are idealizations, we do not expect real organisms to move in perfect Lévy walks, just as we do not expect them to perform true Brownian motion. So how can we argue that organisms perform Lévy walks?

In Section 9.3, we discussed how ballistic motion and normal diffusion described by Gaussian propagators represent the two extremes of Lévy flights and walks. Ballistic motion corresponds to the limit of the Lévy index tending to zero, $\alpha \to 0$. Normal diffusion corresponds to $\alpha = 2$. Lévy walks and flights correspond to intermediate values of α. What are the most important characteristics of this intermediate range? The answer can be found in Chapter 3, i.e., fat tails in the propagator and superdiffusive Hurst exponents $H > 1/2$.

We can now formulate the Lévy flight foraging (LFF) hypothesis more precisely: *superdiffusive motion governed by fat-tailed propagators optimizes encounter rates under specific (but common) circumstances; hence some species must have evolved mechanisms that exploit these properties of Lévy walks.*

Unlike Darwin's theory of natural selection or the laws of thermodynamics, which are universally valid, we have no theoretical basis for expecting that all organisms must have evolved adaptations for performing Lévy flights. Drawing a parallel with eyesight as an adaptation, we know that vision optimizes the success of encounters under specific circumstances, e.g., predation and mating in illuminated environments. Hence, although we know that the ability to detect and transduce photons evolved at least once, we have no basis for assuming that vision is a universal adaptation. Thus, when it is dark, eyes are useless and, similarly, when targets are abundant, Lévy flight foraging is useless. In Section 10.2, we show that the advantages conferred by Lévy flight foraging disappear almost completely under scenarios of high target density [19, 22, 393].

More generally, we cannot expect that large-scale behavior will always reflect underlying resource distributions and other environmental factors. For example, we have no reason to expect that resource distribution will determine animal movement, except for low resource densities. Roshier *et al.* [318] studied movement responses of the gray teal (*Anas gracilis*) in agricultural and desert landscapes. As expected, the movement of the birds was less tortuous in the desert, associated with scarce resources. Remarkably, however, the movement of the birds in the agricultural basin exhibited both high and low tortuosity. Such variation would not occur if resource distribution *uniquely* determined the tortuosity.

The physics of foraging is nontrivial and interesting especially when target densities are low. Indeed, except for low densities (e.g., scarce food in deserts), the target or resource distribution does not appear to play a determining role. Instead, other factors (e.g., internal dynamics) may become more important. For instance, fish [68, 350, 352], insects [197, 348], and mammals (marsupials) [119] have been shown to change their diffusiveness in scarce conditions by increasing movement activity, possibly as an internal neurophysiological response to the low availability of resources. This change in activity supports the idea of Lévy flight foraging as a *built-in* evolutionary adaptation as opposed to an *emergent* property (see Chapter 13).

The LFF hypothesis also assumes Euclidean geometry because geometric and similar constraints are imposed on the search process [234]. For highly constrained searches, where the geometry of the search paths must satisfy species-specific restrictions, the behavior will not be identical or similar to that of unconstrained searches. When termites, for example, excavate tunnels to search for food [202, 203], they create an elaborate primary and secondary tunnel system in three dimensions. In contrast, zebras on the savanna move on a two-dimensional surface. It is entirely plausible that such geometric differences in environment may lead to corresponding differences in movement.

The preceding caveats indicate that we should not expect a single universal law to apply in every situation because the concepts of reductionism and universality, although extremely powerful, suffer from inherent limitations. Invoking again the analogy with vision, consider the general statement that humans do not possess the ability to see images with their ears, i.e., echolocation. This is true for 99.99% of the human population, but it is not universally true. Blind people who have taught themselves such echolocation skills have been documented since 1749. For example, Daniel Kish, who lost his eyesight as a child, can see images of his environment by making clicking sounds with his mouth [184]. Notwithstanding such counterexamples and special cases, nobody would seriously argue that humans possess echolocation abilities. This example shows why physicists apply reductionism to understand complexity: it is easier to start with general limiting models, later adding layers of specifics to address counterexamples, than to attempt the reverse.

Moreover, in addition to the practical advantages of the reductionist approach, empirical observations suggest that it is both valid and robust. Wells *et al.* [405, 406] studied the long-tailed giant rat (*Leopoldamys sabanus*) in logged and unlogged rain forests in Borneo and found that the two settings did not differ with respect to features that are important to movement patterns. If specific (non-universal) aspects of the complexity of the forests were important, then we would expect significant differences in rat movement in the two kinds of forests. This insensitivity suggests that the statistical physics approach of studying limiting models that ignore the effects of, e.g., complex environments may not in fact represent excessive reductionism. Such limiting models may help to quantify important features of the dynamics and the underlying spatiotemporal patterns in the kinetics of animal movement.

In Chapter 2, we discussed "brainless" versus brained random walkers. Returning to that example, it makes much more sense to add, incrementally, the effects of a brain to a limiting "brainless" model than to start with a model of a fully brained animal and attempt the reverse. Indeed, it is much easier to add layers of complexity to a limiting model than to simplify a model with 10^2 free parameters. The LFF hypothesis should be seen in this context: it is a limiting model.

10.2 Analytical and numerical results

Destructive and nondestructive foraging

We review here the analytic solution [393] of the model, defined in Chapter 9, for $p(\ell_j)$ given by Equation (10.1). The result, showing that Lévy flights can optimize random searches under certain conditions, attracted wide attention in the scientific community and helped to overturn the conventional wisdom about the relevance of Lévy processes in biological interactions. It motivated many further studies, both theoretical and experimental [295, 393, 397].

Let λ be the mean free path of the forager between successive target sites randomly distributed. In a two-dimensional landscape, $\lambda \equiv (2r_v\rho)^{-1}$, where ρ is the target site area density. The mean flight distance for the model introducted in Section 9.2 is thus

$$\langle \ell \rangle \approx \frac{\int_{r_v}^{\lambda} dx\, x^{1-\mu} + \lambda \int_{\lambda}^{\infty} x^{-\mu} dx}{\int_{r_v}^{\infty} x^{-\mu} dx} = \left(\frac{\mu-1}{2-\mu}\right)\left(\frac{\lambda^{2-\mu}-r_v^{2-\mu}}{r_v^{1-\mu}}\right) + \frac{\lambda^{2-\mu}}{r_v^{1-\mu}}. \quad (10.2)$$

The second term is an approximation because it assumes that the distances between successive sites are at most identically equal to λ so that there are no flights longer than λ. A new target site is always encountered a maximum distance λ away from the previous target site, effectively resulting in a *truncated* Lévy distribution [221]. As we discuss below, this truncation may play an important role in the context of experimental studies.

We can define the search efficiency function $\eta(\mu)$ to be the ratio of the number of target sites visited to the total distance traversed by the forager. Since this distance is equal to the product of the total number of flights and the mean flight length $\langle \ell \rangle$, therefore,

$$\eta = \frac{1}{N\langle \ell \rangle}, \quad (10.3)$$

where N is the mean number of flights taken by a Lévy forager in order to find two successive target sites.

Consider first the case of *destructive search*, when the target site is eaten or destroyed by the searching animal and becomes unavailable in subsequent flights. The mean number of flights N_d taken to cover an average distance λ between two successive target sites scales as

$$N_d \sim (\lambda/r_v)^{\mu-1}, \quad (10.4)$$

for $1 < \mu \leq 3$, whereas $N_d \sim (\lambda/r_v)^2$ for $\mu > 3$ (normal diffusion) [393]. The exponent $\mu - 1$ is actually the Lévy index α.

Let us first assume the common scenario in which the target sites are *sparsely* distributed, defined by $\lambda \gg r_v$. Substituting Equations (10.2) and (10.4) into Equation (10.3), one finds that the mean efficiency η has no maximum, with lower values of μ leading to more efficient searches. Actually, when $\mu = 1 + \epsilon$ with $\epsilon \to 0^+$, the fraction of flights with $\ell_j < \lambda$ becomes negligible, and the forager effectively moves along straight lines until it detects a target site.

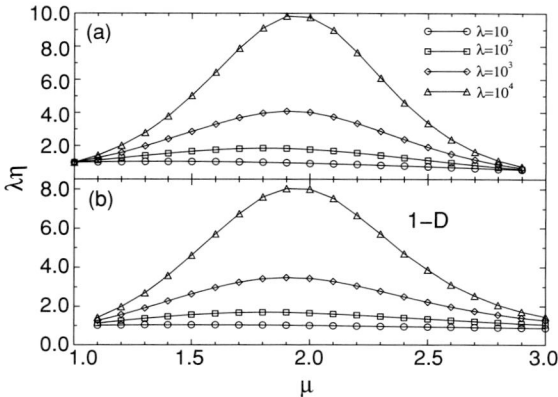

Figure 10.1 The product of the mean free path λ and the foraging efficiency η vs. the Lévy parameter μ in one-dimensional nondestructive searches for different λ, found (a) analytically and (b) from numerical simulations [393]. Note the peak efficiency near $\mu = 2$.

Consider next the case of *nondestructive search* for sparsely distributed target sites. Because previously visited sites can be revisited, the mean number N_d of flights between successive target sites in Equation (10.4) overestimates the true number N_n. Suppose that $N_n \sim N_d^{1/2}$ holds generally; it follows that

$$N_n \sim (\lambda/r_v)^{(\mu-1)/2}, \tag{10.5}$$

for $1 < \mu \leq 3$, whereas $N_n \sim \lambda/r_v$ for $\mu > 3$ (again, Brownian, normal diffusion). Indeed, it has been proven that Equation (10.5) is, in fact, rigorous [62, 63]. This result has been found to become increasingly accurate as (λ/r_v) increases. Note that if $\lambda \gg r_v$, then $N_d \gg N_n$. Substituting Equations (10.2) and (10.5) into Equation (10.3), one sees that the optimal efficiency η is achieved at

$$\mu_{\text{opt}} = 2 - \delta, \tag{10.6}$$

where $\delta \sim 1/[\ln(\lambda/r_v)]^2$ is small. Therefore, in the absence of *a priori* knowledge about the locations of the randomly distributed target sites, an optimal strategy for a forager is to choose $\mu_{\text{opt}} = 2$ when λ/r_v is large (low concentration of targets) but not known exactly. Figures 10.1 and 10.2 compare the analytical results with numerical simulations.

There is a special value of the power law exponent μ that optimizes random nondestructive searches, i.e., $\mu = 2$. However, the existence of such an optimal power law exponent is not surprising. A finite optimal power law exponent also arises, for example, in stochastic resonance [123]. What is surprising is that the

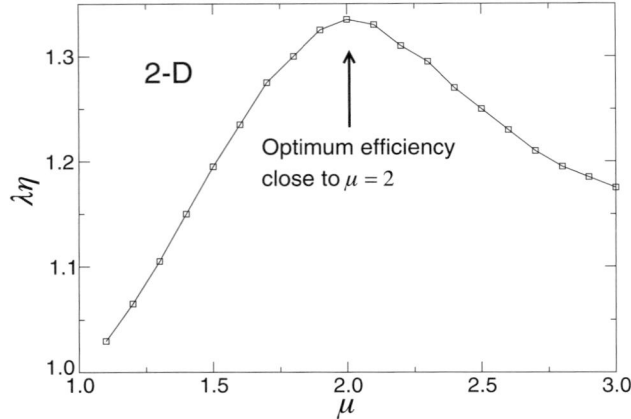

Figure 10.2 The product of the mean free path λ and the foraging efficiency η vs. the Lévy parameter μ for two-dimensional nondestructive searches obtained from numerical simulations with $\lambda = 5000$ ($r_v = 1$) [393]. The finding of peak efficiencies for $\mu = 2$ motivated many further studies.

$\mu = 2$ result is robust when many kinds of perturbations or noise are added to the random search problem (see further below).

Note that even if $p(\ell)$ is a nontruncated power law, the distribution of actual jump lengths will never be identical to $p(\ell)$ for any finite target density because of truncation. When a target is found, often the step ℓ_j will be aborted, leading to a truncated step. For this reason, all the moments of the actual length distributions will be finite, and all moments will be smaller than the moments of $p(\ell)$.

Note that if λ is comparable to or not much larger than r_v, then most flights become truncated no matter what the value of μ, so Lévy flight foraging becomes effectively indistinguishable from Brownian foraging and so loses all adaptive value. Lévy flight foraging in high-target-density scenarios is as useful as eyesight in total darkness.

Generalization to regenerative targets

We can further generalize this result to the intermediate *regenerative* foraging regime. We can associate with both the nondestructive and destructive cases of the random search problems a typical target-regeneration delay time τ. The cases $\tau \to 0$ and $\tau \to \infty$ correspond to the nondestructive and destructive cases, respectively [294, 320]. The equations for N_d and N_n then become special cases of a more general equation for N_r for regenerative targets. Note that N_r must increase monotonically with τ (for fixed μ and λ/r_v). Thus, for a general $\tau \geq 0$, we can define an arbitrary function $\Gamma(\tau)$ where $\Gamma \to 2$ as $\tau \to 0$ and $\Gamma \to 1$ as $\tau \to \infty$

such that $\Gamma(\tau)$ decreases monotonically with τ and $N_r \sim \lambda^{(\mu-1)/\Gamma}$. Maximization of $\eta(\mu)$ implies

$$\{1 - (\lambda/r_v)^\delta\}\Gamma + \{(\Gamma - 1)(\lambda/r_v)^\delta + 1 - \delta\}\delta \ln[\lambda/r_v] = 0, \tag{10.7}$$

where $\mu_{\text{opt}} = 2 - \delta, 0 \leq \delta < 1$. For $\Gamma > \Gamma_{\min} = \ln[\lambda/r_v]/(\ln[\lambda/r_v]-1+(\lambda/r_v)^{-1})$ and fixed λ/r_v, the solution of Equation (10.7) leads to a single peak in η at $1 < \mu_{\text{opt}}(\tau) \leq 2$, so μ_{opt} decreases with respect to the nondestructive case. For $\Gamma < \Gamma_{\min}$, η presents a maximum at $\mu = 1^+$, decreasing monotonically for $\mu > 1$, similarly to the destructive case, where $\Gamma_{\min}$ relates to the revisitability threshold for there to be a peak in $\eta(\mu)$ for a physical value of μ.

All the preceding analytical results [320, 393] have special significance. They imply that if the necessary conditions of the theory hold true, then (truncated) Lévy walks with $\mu \approx \mu_{\text{opt}}$ optimize random searches in sparse environments when target sites are randomly distributed. These results are particularly general, robust, and "coercive."

10.3 Discrete versus continuous media

Random search on lattices and networks

Studies by Santos *et al.* [321, 323] found similar but not identical behavior for noncontinuous discrete media. The specific topology of the searching landscape is a fundamental factor; hence random searches in discrete spaces have features that differ from searches in continuous (Euclidean) environments.

In many potential technological applications, such as computer searches on the Internet, the search is performed in discrete or digital spaces. This is also the case in some neural networks [258] as well as in automated computer searches of registers in high-capacity databases [286, 381]. There are also more biologically oriented examples [231]. Gene networks [344] can form what is called a large-world network [270]. In this case, a random search can be associated with the random mutation problem [264]: a mechanism that can promote the evolution of quasi-species toward local maximums in the fitness landscape (the genotype sequence space). Although the space is huge, it is discrete. Finally, we can cite some quotidian situations, such as looking randomly for a grocery store in the streets of a small or medium-sized town [73], which is effectively a search process on a network structure (i.e., the streets).

There are reasons for studying random searches over discrete structures. Such an analysis is interesting in its own right and may have practical applications. Because continuous and enumerable landscapes can, in certain instances, share a number of similar properties, we may be able to infer useful results for the continuous scenario (e.g., animal foraging) by addressing the less complex discrete lattice scenario.

That being the case, what are the key features determining the transport characteristics of a lattice and thus influencing the outcomes and properties of a random search on it? Many studies [1, 6, 399] have confirmed the hypothesis that such features are related to the topological characteristics of the lattice, particularly the site connectivity. Thus, for structures that are locally connected and rich in clusters, with only short-range links (large-world structures), the searches resemble the continuous Euclidean case. In contrast, small-world networks are globally connected [15]. They possess power law link distance distributions so that rare long-range connections act as shortcuts, reducing the number of links necessary to cross the whole system. Therefore, if not distances but rather the number of steps is the most important factor in the optimization process of a random search (precisely the case in digital spaces), then the searching optimization on such structures is not a critical issue. In fact, small-world networks already have as an intrinsic property a power law distribution of lengths for the bonds, and a Brownian random search on a small-world network is already a kind of Lévy walk (or Lévy flight), but with an exponent determined by the distribution of bond lengths. It is an open question whether a Lévy walk produces better search efficiency than a Brownian search on a small-world network (see Chapter 14).

In contrast to small-world networks, large-world lattices, usually associated with less efficient systems for transport, are more sensitive to the method used for locating randomly distributed target sites, in analogy with the continuous case. Note that there is an important reason to expect that these network Lévy flight searches will result in the best random search strategies: a metric similar to continuous spaces. Indeed, in both situations, the farther away a target site, the longer the distance traveled to reach it. This is not true for (nonscaled) small-world lattices in which a single bond can connect two distant nodes. With these similarities in mind, we summarize some general aspects associated with the random search problem in large-world lattices. More detailed technical discussions are available in the original studies [321, 323].

As with continuous topologies, several factors determine the outcome of a random search in discrete topologies (Figure 10.3). Among the most relevant are connectivity, boundary conditions, target density, and number and position of defects. In the following, we address each one of these aspects for large-world structures in regular and isotropic finite lattices, characterized by lattice parameter s, coordination number k (which is the number of different directions that the search can follow from a given node), and typical size $L = N_0^{1/d} s$, with d being the spatial dimension and N_0 the total number of nodes.

In such structures, defects can be created by randomly eliminating a certain fraction of nodes (Figure 10.3). If N_0 is the initial number of nodes in a perfect lattice, then a *fragmentation coefficient* is defined by $\chi = N_D/N_0$, where N_D is the

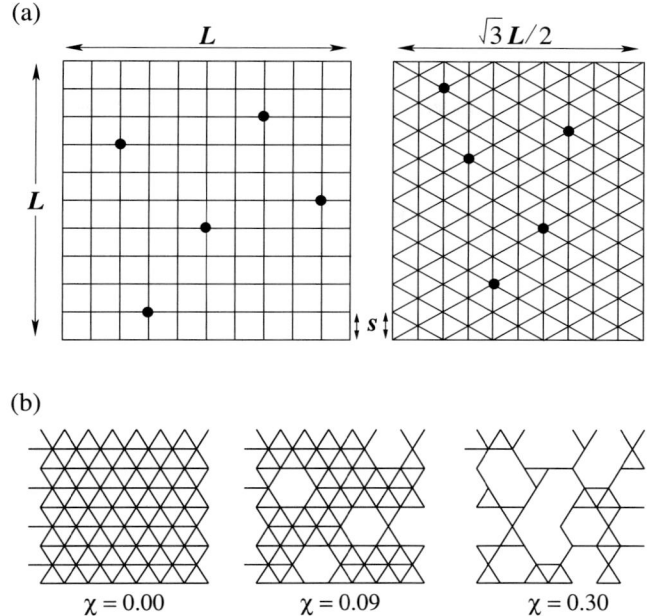

Figure 10.3 (a) Examples of square and triangular lattices. The dots at the nodes represent randomly distributed target sites. (b) Different defective lattices.

total number of nodes removed or destroyed ($\chi = 1$ implies an empty space – total dilution or complete destruction of the lattice).

Regarding the boundary conditions (BCs), we first note that they play a significant role in the search optimization process only when the searcher reaches the lattice borders (or equivalently, travels a distance equal to the system size) with a certain frequency. Hence the BCs are in fact relevant only if the target site density and number of defects are low. Otherwise, the many search path truncations will decrease the number of large excursions comparable to the lattice size. There are many possible choices for the boundary conditions. Consider a totally biased, non-searching walker moving in the straight lines allowed by the lattice and according to the rules imposed by the BCs at the lattice borders. The BCs in this case can be characterized in terms of the kind of trajectories (described earlier) they generate. The BCs (depending on the lattice topology) can lead to (1) a looplike trajectory that returns to itself after traveling a distance L, (2) a trajectory that scans only a fraction of the lattice, or (3) a trajectory that covers the entire lattice. In a hypercube lattice in any number of dimensions, for example, periodic boundary conditions result only in type 1, whereas for a triangular lattice in 2D [323], it gives rise to both types 1 and 3. Any BC that leads to type 3 will give rise to ergodic behavior for infinitely long trajectories, a point we discuss further below.

The search rule dynamics are similar to those for the continuous case previously discussed. The only differences are that (1) the directions are only those allowed by the lattice structure, and the movement takes place in a straight line of bonds; (2) if a defect exists along the direction of step j (within distance ℓ_j), and no target site is found before reaching it, then the step is truncated at the node immediately before the defect, a new direction and step length are chosen, and the process resumes; and (3) an efficiency function η, defined as either the averaged distance traveled per target or the neat energy gain per target, is an appropriate quantity to analyze the different strategies.

Results obtained in the literature for the large-world lattices described earlier can be summarized as follows. Exactly as in the continuous case, the efficiency function becomes less and less sensitive to the Lévy exponent μ for increasing target densities. Also, the boundary conditions are not relevant to the search outcome. In the absence of defects, for low target density and a nondestructive search, $\mu_{opt} \approx 2$ is the optimal exponent. Also, in this regime, BC resulting in more ergodic trajectories (i.e., those visiting a larger fraction of the lattice) lead to a general increase of the efficiency function η for $\mu < 2$. For $\mu > 2$, long walks are rare, so the borders are seldom visited, and the BCs do not significantly affect η.

In the regime of destructive search and low density, there is a remarkable difference between the continuous and discrete cases. When the BCs generate trajectories of type 1 or 2, then extremely long walks decrease the efficiency because a long step can become a long loop, without finding any target site. Such behavior does not favor ballistic strategies, which maximize efficiency in destructive random searches when the environment is continuous but which can lead to recurring infinite loops, depending on the BCs. When such conditions prevail, a destructive search is optimized with exponents in the range $1 < \mu \leq 2$ and is thus similar to the nondestructive case.

As expected, an increase in the coordination number k makes the efficiency curves tend to the same shape as those for continuous landscapes. For a (low) fixed target density, larger values of k lead to a larger efficiency η for the same μ. This is the case because a larger number of possible directions at each step, given by k, decreases the importance of the trajectories of types 1 and 2, both of which have lower scanning power on the lattice (because these trajectories cannot cover the whole space).

If the number of defects is too high, all steps are truncated to small displacements, i.e., of the order of the free mean path for defects. Thus, for choosing a particular optimization strategy, setting the value of μ no longer makes any sense; i.e. it is now irrelevant. Assuming that the defect concentration is sufficiently small, the dynamics of flight truncation due to defects are similar to that of normal target sites. The only fundamental difference is that when a flight is truncated by a defect, the searcher does not gain any resource (e.g., food). Such truncation contributes negatively to the global efficiency, the defect acting as a fake target site. Thus, due to the defects, the efficiency function should decrease by a factor weakly dependent on μ and, in a first approximation, we expect a global diminishing of the η curve without any important qualitative change in its shape. Note that in a mean field approximation, the presence of defects merely promotes a rescaling of the target site density. This qualitative reasoning has been verified numerically and analytically [323].

Finally, what is the threshold for the fragmentation coefficient χ such that random search optimization is no longer possible? How robust is a given lattice to defects such that it still allows efficient searching? These are still open questions, but their answers will surely be closely related to percolation problems.

Efficient navigation in a small world

Some of the preceding ideas are connected to the problem of locating small chains between points in a network. Consider the famous problem of finding the shortest path between the sender and the recipient of a message, given only local information. The topic became popularized in the phrase "six degrees of separation."

On a lattice or network, we define one site to be the receiver and another the sender of the message. The goal is to determine the expected delivery time T, i.e., the average number of steps necessary for the information to travel from the sender to the receiver. In small-world networks, unlike regular lattices, there is always a short trajectory linking two sites. Therefore, in principle, T should be small in this case, but how can we find an efficient algorithm for calculating these short paths?

It was recently discovered that, in special small-word lattices, this problem is solvable [187]. Assume a 2-D ($d = 2$) regular square lattice with additional long-range links that connect distant nodes. These long-range links are selected with a probability proportional to $1/r^\alpha$, where r is the distance (along the lattice) between the two sites. A rapid delivery time is obtained when $\alpha = 2$ from the so-called greedy algorithm: at site s, choose to go to the site connected to s that brings you as close as possible (in lattice distance) to the target.

Although this inverse square law is the same as that for random searches, a remarkable coincidence, the results do not agree in other spatial dimensions. In general, the optimal α is equal to the dimensionality of the lattice. In contrast, for the random search problem discussed earlier, the optimal solution is $\mu = 2$ always, regardless of dimensionality. In both cases, however, a key factor seems to be the compromise between the costs and the benefits of long-range versus short-range (i.e., local) "interactions."

10.4 Energy and entropy

Energy and entropy are central concepts in statistical physics. They are also central to foraging, but in a different context. Energy is important because locomotion has an associated energy cost. Entropy is associated with environmental disorder and information availability. In Section 10.2, we did not take either concept into account.

Energy

The statistical efficiency η in Equation (10.3) can be expressed as the reciprocal of the average distance traveled between successive targets. This definition does not take into consideration energy costs and gains. If the targets represent food, as opposed to, e.g., sexual partners, then they have an energy (i.e., caloric) content. There is also an energy cost for locomotion. The simplest assumption is that the energy cost of locomotion is linear in the distance traveled. This is an approximation for the same reason that fuel efficiency of motor vehicles is lower for city streets than for major roads. A more realistic assumption is that the energy cost is a monotonically increasing function of the distance traveled. We can now define an energy efficiency η_E that is nonlinearly related to the statistical efficiency,

$$\eta_E = \frac{\langle E \rangle}{\langle L_t \rangle}, \tag{10.8}$$

where $\langle E \rangle$ is the average energy gained during a specified time interval and $\langle L_t \rangle$ is the total distance traveled, which is approximately the product of the average distance traveled $\langle L_2 \rangle$ between two successive targets and the number N_t of targets found. If each target provides an energy gain of ϵ, and if $f(\langle L_2 \rangle)$ is the monotonic energy cost discussed previously, then we can write

$$\langle E \rangle = (\epsilon - f)N_t. \tag{10.9}$$

Noting that $N\langle \ell \rangle$ in Equation (10.3) equals $\langle L_2 \rangle$, we can write $\eta = 1/\langle L_2 \rangle$, from which it follows that

$$\eta_E = \eta(\epsilon - f). \tag{10.10}$$

If energy is taken into account via the preceding equation, the results of Section 10.2 remain qualitatively unchanged [294, 320]. Specifically, the extrema in η (relative to μ) remain extrema of η_E. That the statistical efficiency η can be used to obtain (exactly) the strategy that optimizes the energy efficiency η_E is yet another example of the predictive power of statistical physics. However, despite the fact that functions η and η_E are formally maximized at the same points, this maximum may not be accessible to η_E because the maximum must also satisfy the constraint $E \geq 0$. A negative average rate of energy gain corresponds to a starving or dying animal.

Entropy, information, and patchiness

Entropy and information are related (e.g., Shannon entropy). Many organisms have nervous systems that learn new behaviors when faced with new information. Information about environmental disorder can be extracted from chemical gradients (e.g., odors). Another source of information comes from memory of the terrain or of the environment in general. Because there are too many possibilities, we do not yet have a comprehensive theory of how information and entropy may interact and modify foraging patterns. Nevertheless, we can once again start with limiting cases.

One type of information pertains not to target position but to target velocity. If the targets can diffuse, we will obtain destructive foraging results [115] similar to those for fixed targets. Moving targets differ from fixed targets in that the searcher can remain still and wait for the target to find the searcher. Spiders, for example, wait for their prey to come to them. However, this passive approach is not always efficient [19].

Organisms with access to information (e.g., learning) that modifies their behavior by introducing short-range correlations in their motion can be identified when there is a well-defined correlation time. Because the correlations and the memory effects become negligible (exponentially) at timescales very much longer than the correlation time, we can show that they do not change the results of Section 10.2 for sufficiently low target densities [90, 394]; however, whether or not such very low target densities are realistic is a separate question.

When there is short-range correlated disorder in the environment, the targets are distributed in an orderly manner at short spatial scales but not long ones. If the correlation distance is smaller or comparable to the radius of vision, then the target density is large, and Lévy flights are not advantageous. On the other hand, if the correlation distance is much larger than the radius of vision, then the searcher's local environment is essentially ordered and predictable, e.g., crystalline. A stochastic strategy is not necessary in either scenario. Certain types of chemotactic movement may indeed not require stochasticity [272]. Such scenarios, however, are neither common nor always realistic. Even bacterial chemotaxis has nonnegligible stochasticity [189]. Specifically, certain bacteria seem to have noise generators in their chemotaxis networks and these additional fluctuations in signal transduction appear to be a selected property [189]. In other words, the noise is there by "design."

A more interesting and realistic case concerns patchy environments, i.e., those in which there are randomly distributed dense target patches with few or no targets outside the patches. Such heterogeneous landscapes are biologically important.

Targets are not uniformly distributed and are usually clustered [401] in (possibly hierarchical [111]) patches. A hierarchical patch structure suggests a fractal target distribution [315]. There is considerable evidence for the existence of hierarchical patch structure [113]. Johnson *et al.* [168] studied patchiness on multiple spatial scales in the context of landscape ecology. Given the high incidence of patchiness in target distributions, it is conceivable that adaptations might have evolved to exploit them [23]. In this context, quantifying how animals respond to local conditions by taking into account their internal states can reveal [122] how they respond to spatial heterogeneity at different spatial scales. Herbivores of widely varying body mass respond to heterogeneity in a remarkably similar manner [155], for example. Hierarchically structured patches also have, to some extent, scale-free or fractal properties, which we expect to be reflected in foraging strategies. Fauchald and Tveraa [112] studied Antarctic petrels, seabirds that forage in a hierarchical patch system. They found results indicating a nested search strategy, exactly as we would expect. Sims *et al.* [347] discussed complex landscapes and fractal prey distributions in similar contexts. Fractal target distributions were one of the original motivations [390] for studying Lévy flight foraging.

There are more complex landscapes that lie beyond the scope of this book. Consider, for example, a rainforest canopy where the local microenvironment at low heights is effectively disconnected or fragmented, while at the top, there is a single connected landscape that spans the entire forest. Sole *et al.* [355] found that long-range dispersal strategies are more useful to species that do not have access to the higher, fully connected region of the canopy.

What happens if there is not one but many searchers? When there are a number $\mathcal{N}$ of walkers searching for targets, then not only must each searcher not oversample the sites it has previously visited, it also must not oversample the sites visited by the other $\mathcal{N} - 1$ walkers. Oversampling becomes even more important, and the optimal strategy approaches ballistic motion as $\mathcal{N} \to \infty$, as argued by Reynolds [309] (see also [36, 37, 413]). On the other hand, if the $\mathcal{N}$ walkers must remain close to each other in space to retain a herd or group structure, the problem becomes even richer. Herds and groups of animals can only remain stable if there is a condensation in both position space and velocity space, i.e., the organisms not only must remain close to each other in position but must also move at similar velocities. It has been shown that a possible way to reconcile large search efficiencies while keeping a group structure is through Lévy walk strategies [324].

In summary, energetic and entropic corrections to the purely statistical considerations of Section 10.2 may lead to qualitative changes in the effectiveness of Lévy flight foraging, although the $\mu \approx 2$ solution remains optimal, or nearly so, in a variety of situations. We close this chapter by noting that there may be advantages to ignoring certain kinds of information. In some circumstances, random search strategies paradoxically outperform chemotaxis [313]. This counterintuitive effect makes sense if one takes into account that the information available may be *misinformation*, i.e., may not be the information needed to optimize the search. Such misinformation could arise not only due to counteradaptations in the targets (e.g., foul-smelling odors emitted by prey) but also because information collection and processing are always error prone.

A similar phenomenon sometimes is found in financial markets, where random buying and selling of assets can financially outperform a sophisticated investment strategy that takes into account many financial indicators. Information has an associated cost not only in finance but also in foraging.

11

Other search models

The results discussed in Chapter 10 have inspired a renewed interest in fundamental questions relating to random searches. We have seen that Lévy flights have scale-free properties such that there is no unique characteristic scale in the random walk flight length (or step length) distribution $p(\ell)$. In contrast, Wiener noise, unlike Lévy processes, has a well-defined characteristic scale because all moments are finite. Are the high search efficiencies of Lévy flight foraging due to the multiple scales or, equivalently, to the scale-free properties? How many scales would be sufficient to guarantee high encounter rates and search efficiencies? Perhaps scale-free properties are not needed after all, and a few scales would be sufficient. In this chapter, we review search models that contain free parameters embedding characteristic scales.

11.1 Correlated random walks with a single scale

The most natural and obvious choice for the fewest number of characteristic scales is one. Correlated random walks (CRWs) appeared the study of ecology when short- and medium-scaled animal movement data were analyzed. CRWs have a single characteristic scale – a correlation length or time that can be quantified via sinuosity. Experiments with ants, beetles, and butterflies were performed in 15 to 20 square meter arenas as well as in their natural environments (and usually lasted fewer than 45 minutes). From these studies, ecologists promptly became aware of the necessity of adding directional persistence to pure random walks to reproduce realistic animal movements [21, 42, 173].

More recently, the mathematical properties of CRWs were used to explore the link between individual animal movements and population-level spatial patterns [377, 378]. Further studies have considered the relative straightness of the CRW (i.e., degree of directionality [146] or sinuosity [27, 41, 43]) as relevant properties characterizing animal movement [21].

100

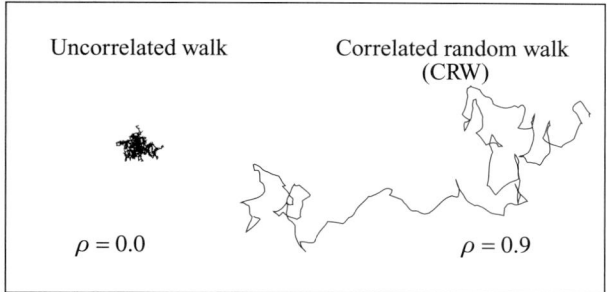

Figure 11.1 Correlated random walks (CRWs) are different from uncorrelated random walks due to directional persistence. The turning angles between successive random walk steps are not uniformly distributed for CRWs. Instead, typically, they are chosen so that the most probable turning angle is zero, i.e., the random walker tends to retain the original direction. The plot shows an uncorrelated random walk and a CRW of the same total length. (Figure by M. L. Felisberto.)

Correlations are introduced into the random walk as follows. In two dimensions, typically, two random walk step vectors differ only in their angular directions. The *turning angles* θ_j between successive step vectors $\mathbf{r}_j$ and $\mathbf{r}_{j+1}$ are usually chosen from a symmetric distribution so that we can define the mean resultant,

$$\rho = \langle \cos(\theta) \rangle. \tag{11.1}$$

We show in Appendix A that this one-step Markov process has a single characteristic correlation time proportional to $-1/\ln \rho$. At scales much larger than the correlation time, such CRWs behave diffusively. CRWs are equivalent to a small-angle scattering process. Figure 11.1 illustrates the main features of CRWs.

More generally, the direction of random walk step j may bear a correlation to the previous n steps (with the most common value being $n = 1$). Because the next random walk step is independent of all steps k, where $k < j - n$, such walks are n-step Markov processes and cannot sustain long-range power law correlations. Instead, they possess a well-defined correlation time [396] – hence the term *short-range correlations*.

Indeed, CRWs and other Markov processes become effectively uncorrelated at timescales much longer than the correlation time. Therefore they necessarily lead to normal diffusion at long timescales [23, 396]. Although the mean squared displacement grows quadratically at short times, it nevertheless grows linearly at times much longer than the correlation time (Figure 11.2). The behavior is not unlike that of the telegrapher's equation (see Chapter 3). The propagator for CRWs

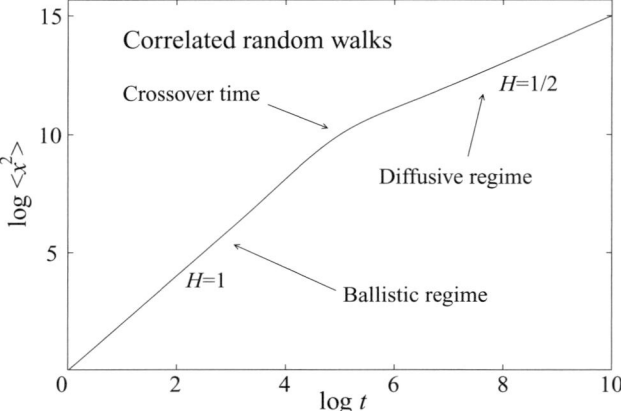

Figure 11.2 Sketch of the mean squared displacement of CRWs. At short timescales, the directional persistence of CRWs leads to ballistic behavior. However, because CRWs are Markov processes, the two-point autocorrelation function of the velocity vector decays exponentially. The inverse decay constant represents the mean life or correlation time for the direction persistence. At timescales very much larger than this characteristic time, the behavior crosses over from ballistic to diffusive.

converges to a Gaussian at long times. For this reason, CRWs cannot explain fat tails, yet fat tails are observed in dispersal and movement data.

CRW models have been studied in the context of biological mechanisms that link information processing directly to the sinuosity parameter ρ. However, such microscopically detailed models are not the main topic of this book.

Lévy-modulated CRWs and correlated Lévy walks

We mention two hybrid models that incorporate aspects of both Lévy walks and CRWs. Bartumeus *et al.* [21] proposed Lévy-modulated correlated random walks (LMCRWs), which are equivalent to concatenated CRW stretches whose lengths are asymptotically power law distributed. A model very similar to LMCRWs with nonconstant sinuosity parameterized along the trajectory by the Lévy walk flight length ℓ_j has been shown to be genuinely superdiffusive [396]. It is also possible to create hybrid models without concatenation. Felisberto *et al.* [116] studied correlated Lévy walks (CLWs) – Lévy walks with nonuniformly distributed turning angles (Figure 11.3). Narrow turning angle distributions lead to additional persistence at short timescales, leading in turn to ballistic behavior. The CLW model is Markovian in the step number j index for flight ℓ_j, but it is not Markovian in the actual time t, so it retains superdiffusive properties at arbitrarily large timescales, similarly to uncorrelated (i.e., conventional) Lévy walks.

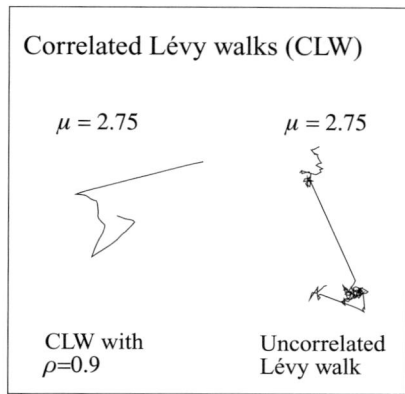

Figure 11.3 Correlated Lévy walks (CLWs) are different from uncorrelated Lévy walks due to directional persistence, in a manner analogous to how CRWs are different from uncorrelated Brownian random walks. CLWs, like CRWs, are ballistic at short times but may become superdiffusive at larger timescales. CLWs are more exotic than either CRWs or Lévy walks, and their properties are still under investigation. (Figure by M. L. Felisberto.)

11.2 Intermittent searches with two scales

CRWs have one characteristic scale. We now examine intermittent searches with two scales. We obtain two characteristic scales by alternating between two modes in a bistable process. This leads to intermittent dynamics – hence the interest in intermittent random search models [30, 32, 33, 34, 253, 254, 268, 269, 339], i.e., those consisting of two kinds of motion, each with its own special characteristic scale.

Figure 11.4 illustrates key features of intermittent searches with two phases, as studied by Bénichou *et al.* [33]. The locally static or diffusive search phases alternate with ballistic relocations, during which detection is typically switched off. There is no detection during relocation, unlike the Lévy flight model where detection is always present.

Besides the absence of searching during ballistic relocations, an important feature of intermittent search models relates to the time or distance distributions for the two phases. In such models, both the diffusive phase and the ballistic relocation times (or distances) have finite moments. One example [33] has an exponentially decaying probability density for the ballistic relocation times.

As with all optimizable models, a global optimum for search efficiency can be found in the free parameter space of intermittent search models. For example, Lévy flights with $\mu = 2$ optimize searches for revisitable targets, yet intermittent searches can be optimum with other values of μ [33]. The alternation of *intensive* search phases and *extensive* relocation phases – when food becomes locally

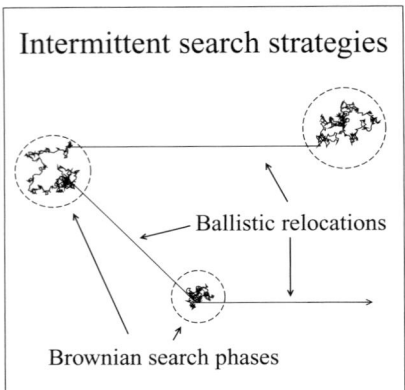

Figure 11.4 Intermittent searches typically consist of alternating phases: a Brownian search phase alternating with ballistic relocations. The ballistic relocations help to reduce oversampling, in a manner similar to how Lévy walks help reduce oversampling.

depleted – leads to a search efficiency that is claimed to be higher than that of the original Lévy model [288] (see also Section 11.2). Properly weighting the switching times leads to an increase in the plasticity of intermittent models, which allows global optimization in diverse situations. For this reason, it has been claimed that intermittent models can have higher search efficiencies than the original Lévy walk models, but this is as expected – the free parameter space for intermittent searches is usually larger (i.e., has higher dimensionality) than for Lévy walk models. By optimizing the switching times, one can find a set of free parameters that will optimally exploit the characteristic scales for the problem. In contrast, for Lévy strategies, one optimizes the search efficiency by changing a single free parameter, the power law exponent μ. A fair comparison of intermittent models with Lévy walk models having the same number of free parameters would give a clearer picture.

For nonrevisitable targets, the optimal value of μ in Lévy searches corresponds to ballistic searches. It is also difficult to compare intermittent searches with ballistic searches because in intermittent search models, the detection is typically switched off during the ballistic phase. This aspect of intermittent search models may not always be biologically realistic. If a species has evolved an adaptation for detecting a particular kind of target (e.g., food), then it makes no sense to evolve a second adaptation to switch off the first adaptation. Males in search of females do not switch off their ability to detect females during large relocations. Nevertheless, the study of intermittent search models has helped us advance our theoretical understanding of search processes [339].

It has been claimed that intermittent search models may represent an evolutionary alternative to Lévy searches [33]. Do intermittent searches correspond to the actual movements of real organisms? Intermittent searches appear to work remarkably well for some species. For example, they may correctly describe the phenomenology of planktivorous fish and ground-foraging birds [304]. Atlantic bluefin tuna (*Thunnus thynnus*) also appear to switch between two search modes [262]. Houseflies seem to switch their style of motion from low-dimensional regular patterns to high-dimensional disordered patterns [368] in a manner that suggests neither Lévy flights nor Lévy walks. Elk may also adopt intermittent searches [252].

A study by Oshanin *et al.* [269] provides insight into why intermittent searches are efficient. They study a search model in which the searcher can move in one of two ways on a lattice. The searcher either moves to a neighboring site with probability p or, with probability $1 - p$ relocates ballistically by moving off-lattice a distance L in a random direction and then returns to the lattice. During the off-lattice relocation, the searcher cannot interact with (i.e., detect) targets (we follow a slightly different notation from that in [269] to avoid confusion with the notation used elsewhere in this book). Realistically, we assume a fixed, constant velocity for the searcher. A double optimization over p and L leads to the best search strategy. As expected, this double optimization outperforms the single optimization of the original Lévy walk models – the optimal intermittent strategy bests a Cauchy (i.e., $\mu = 2$ Lévy stable) distribution. These intermittent searches are efficient because the singly optimized Cauchy searches take the searcher too far, even when there is no need to go infinitely far. Recall that the first moment of the step length distribution diverges for $\mu = 2$ when there is no truncation cutoff. Recall that in the limiting model of search introduced in Chapter 9, an ultralong flight length becomes truncated when a target comes within the detection radius. Diverging flight lengths thus are not present in that model. In intermittent searches, however, the detection is switched off during long relocations, causing the singly optimized Cauchy strategy to lead to unnecessarily long flights. We might attempt to doubly optimize such off-lattice Lévy searches with respect not only to the Lévy index but also to a truncation or cutoff length L. Such a cutoff would ensure finite moments for the distribution of flight lengths, no matter the value of the Lévy index.

Scale-free intermittent searches

Intermittent searches were originally proposed as an alternative to Lévy flight strategies [33], but it has since been shown that intermittent searches and Lévy searches can be combined [213]. Moreover, each type of model may find relevance in distinct scenarios. In intermittent searches, there are two phases: (1) a Brownian search phase and (2) a ballistic relocation phase. The original intermittent search models assumed that all the moments of the distribution ballistic relocation distances are finite, but due to the central limit theorem, this assumption forces the motion to become diffusive, even in the limit of zero target density. Clearly this is a restrictive and unnecessary assumption.

Lomholt *et al.* [213] weakened this assumption by allowing power law tailed distributions of the ballistic relocation times. They demonstrated analytically and numerically, in one dimension, that when relocation times are Lévy distributed, the search process significantly outperforms the previously investigated case of exponentially distributed relocation times. The addition of the Lévy ingredient to the intermittent model reduces oversampling, thereby further optimizing the intermittent search strategy in the biologically important case of rare targets.

Their results are consistent with what we discussed in Chapter 10. For sufficiently low target densities, Brownian intermittent searches lead to normal diffusion and the usual problems of oversampling. In contrast, Lévy intermittent searches of the kind studied by Lomholt *et al.* retain the ability to avoid oversampling for arbitrarily low target densities.

Lomholt *et al.* showed not only that Lévy intermittent searches reduce oversampling but also that such searches are less sensitive to target density, which allows a much smoother adaptation [213] from an evolutionary perspective. This reduced sensitivity to the length scale associated with target density is a direct consequence of the scale-free properties of Lévy flights.

But how and why can Lévy intermittent searches outperform Brownian intermittent searches when we know that Lévy flights take the searcher too far? The answer is twofold. On one hand, the results of Lomholt *et al.* show that only Lévy intermittent searches with $2 < \mu < 3$ (corresponding to Lévy indices $1 < \alpha < 2$) are better than Brownian intermittent searches. These values are precisely those required to avoid diverging mean flight lengths (or times) in the limit of zero target density. On the other hand, Lomholt *et al.* also make clear that when target densities are low, Lévy intermittent searches improve [213], and when targets are infinitely rare, no distance is too far. This is the bottom line for scale-free models.

11.3 A unified approach

CRW searches, intermittent searches, and Lévy search models can all be seen as reaction-diffusion processes. They differ from each other in that the statistical properties of their movements differ. CRWs have a single (possibly time-dependent) characteristic scale: the correlation length or sinuosity. Intermittent models have two or more characteristic scales, e.g., switching times and the upper cutoff for ballistic relocation. Lévy search models have scale-free properties. The three are more alike than different and are neither mutually exclusive nor incompatible. Reynolds [297] has shown that intermittent searches can be understood within the context of an optimal scale-free, Lévy flight searching strategy, and we have already discussed hybrid models of CRWs and Lévy walks.

Our task is not to compare competing theories of optimality because nature is replete with suboptimal adaptations. Consider, for instance, a CRW whose corresponding step-length distribution $p(\ell)$ has a characteristic scale and whose moments are all finite. It will display a degree of directional memory, introduced on the basis of a nonuniform distribution of turning angles [378]. As a Markovian process, for large timescales, it will present normal diffusion, i.e., $H = 1/2$. Hence, for any random search requiring superdiffusion, this type of random walk is not optimal. Yet CRWs have been successfully used to describe the movement of a variety of species. Which random walk models lead to an adequate outcome given the actual environmental, metabolic, and (eventually) social conditions and constraints of the process? The answer could be Lévy walks, intermittent walks, correlated random walks, or even something more exotic (Figure 11.3), depending on which model is adequate in a given, specific search scenario.

Given such considerations, we wish to avoid a Manichaean discussion about whether Lévy flights are a better option than composite random plus ballistic intermittent strategies (and other similar discussions). Many possibilities were perhaps explored to improve stochastic searches throughout evolutionary history. Owing to the huge variability and diversity of biological systems, we can imagine that natural selection has followed distinct – nonuniversal – pathways, often finding good (but not necessarily optimal) solutions, depending on specific circumstances.

Despite the preceding discussion and the complexity of real organisms, it is still remarkable how well Lévy flight patterns account for widely differing scenarios, many of which have been described in Chapter 6. Historically, the idea of scale-free Lévy flight foraging constituted a conceptual advance and inspired a number of further studies, but there is no reason to expect the innovation to cease with *plain* Lévy flights. As a recent example, curved Lévy trajectories have been proposed [396]. There are a number of possibilities for interesting theoretical and experimental studies in the near future that could reveal still more profound and unexpected features of the rich dynamics underlying biological encounters.

Part IV
Finale: A broader context

12

Superdiffusive random searches

Although this book focuses on searches performed by foraging animals and other organisms, the physics underlying foraging overlaps with that of several other kinds of random searches and stochastic optimization problems. What follows is a brief discussion of this interdisciplinary field of research.

12.1 Submarine warfare and operations research

World War II saw the successful application of mathematical analysis to warfare, one of the most famous examples being the detection of enemy submarines ("submarine hunting"). Morse and Kimball [255, 256] (see also [263]) have provided us with some details. Once the German military captured France, the Bay of Biscay ports became the principal operational bases for their U-boats. Most German submarines operating in the Atlantic went out to sea and returned to port through the Bay of Biscay. Starting in 1942, the Royal Air Force (RAF) had enough long-range planes to dedicate several of them to antisubmarine duty. The submarines had to be detected prior to attack. Because the sole mission of these planes was to detect and destroy submarines, their success in the campaign was measured using the number of U-boat sightings made by the aircraft [256].

Morse and Kimball [256, 340] introduced the operational quantity Q, measured in square miles per day:

$$Q = \frac{AC}{NT},$$
(12.1)

where A is the area surveyed, C is the number of contacts in a time T, and N is the number of submarines in A. They also have a theoretical expression for Q:

$$Q = 2Rv,$$
(12.2)

where R is the detection radius and v is the speed of search. The specific techniques of operations research were aimed at optimizing the measured values of Q.

Without going into detail, we note that there is a degree of complexity to the problem that arises due to the adaptation of countermeasures, counter-countermeasures, etc. For example, a sudden drop in Q might indicate the introduction of a new and successful countermeasure. In 1942, the Germans started equipping their submarines with (L-band) radar detectors as a countermeasure to British radar. With the advance warning this provided, the alerted submarines could submerge before being seen by the RAF. Submarine warfare can be "a constant game of tactics and countermeasures based on advances in learning and implementing new technology" [340].

Because the hunt for submarines represents a kind of random search process, could Lévy walks find application in such problems related to operations research? Shlesinger [340] has discussed similarities in many different search problems, spanning the hunt for submarines, predator-prey relations, and even molecular searches. Without explicitly specifying the application to operations research, Shlesinger notes that the Lévy walk beats the Brownian walk for "escaping" large volumes devoid of targets [340], so it is not inconceivable that Lévy walks are already used for military purposes in the "constant game of tactics and countermeasures."

12.2 Enzymatic searches on DNA

When any situation presents many similar pairs of keys and locks, considerable time and energy are required to find the matching pairs. As the number of pairs grows, so does the time needed to find the right key-lock combinations.

Such situations can be found at the chemical level [7, 104, 281]. When a substrate approaches an enzyme, it can only couple or enter the active site of the enzyme and form the enzyme-substrate complex if the substrate matches the enzyme. The search halts only when the appropriate coupling takes place. Searches [107, 108, 159, 212, 238, 246, 249] performed by enzymes on DNA strands for specific sequences are a particularly important example. Indeed, cellular processes provide extraordinary examples of the remarkably complex dynamics of molecular reaction pathways. The problem has many aspects and involves multiple characteristic timescales. There is diffusion in three dimensions (3-D) when the enzyme moves in the medium surrounding the strand. It may or may not perform long jumps during this 3-D diffusion time $t_{3\text{-}D}$ (depending, e.g., on the DNA coiling state [214]). On landing on the DNA chain, the enzyme slides along it, scanning for the correct binding locations and covering a certain number of base pairs. If it does not find the correct match within a certain 1D diffusion time $t_{1\text{-}D}$, the 3-D diffusion resumes. Note that these timescales are affected by such factors as the mechanism for switching between 1-D and 3-D diffusion and the specific chemical interactions between the DNA and protein during the 1-D process [92]. The latter, for example, establishes how the enzyme recognizes a given location as the correct sequence of base pairs [72] and may also determine the sliding velocity.

This rich problem is also the subject of a review by Metzler *et al.* [246] that discusses how DNA looping, DNA knots, and the spontaneous formation of DNA nanobubbles may affect biological processes, e.g., transcription initiation, and how proteins search for their binding sites on a DNA molecule.

Eliazar *et al.* [107, 108] studied a large class of parallel and massively parallel searches in long circular strands for a target site. Their model contained agents that combine local scanning with random relocations on the strand (see Chapter 11 for similar models). They then derived analytically the optimal search strategies. Lomholt *et al.* [212] studied models of searches by proteins of targets on DNA, combining normal diffusion along the chain with Lévy-type diffusion. The superdiffusion along the chain was made possible by such geometrical effects as chain looping.

12.3 Robot foraging

Note that statistical models of random searches, such as the model discussed in Chapter 10, do not assume any particular implementation of searchers and targets. The searcher is usually taken to be a biological organism (or, in the case of DNA searches, a biological enzyme or macromolecule), but the searcher could also be robotic. Because the behavior of random searches is independent of implementation details, successful robotic searches closely resemble biological searches.

Robot foraging, and evolutionary robotics in particular, is a growing field of scientific research [380]. Although robot behavior has traditionally been studied via the microscopic analysis of systems composed of a single or only a few robots, more recently swarms of robots have been studied [232]. Macroscopic robot analysis, in contrast, focuses on averaged quantities. Van Dartel *et al.* [380] analyzed a model of robot foraging and found that successful robots forage in a manner similar to Lévy flight foraging.

12.4 Eye microsaccades

Another fascinating phenomenon is microsaccades [110]. A *saccade* is a quick, simultaneous movement of both eyes that occurs when, e.g., the viewer wants to remain focused on a single spot (visual fixation) or to engage in rapid eye movement [228]. Microsaccades are involuntary, smaller versions of fixational saccades, and their role has been a topic of much debate [227].

A study by Engbert [110] explored the similarities between microsaccades and foraging. Fixation eye movements, according to one hypothesis, represent a search process. Engbert examined the distribution of microsaccades and found a power law exponent $\mu \approx 4.41$. This value for the exponent guarantees finite second moments, thus excluding Lévy processes, but it is also much fatter (i.e., leptokurtic) than a

Gaussian. Brockmann and Geisel [53] showed that human visual scanpaths will be geometrically similar to Lévy flights if one assumes that the visual system minimizes the typical time needed to process a visual scene.

A definitive understanding of the utility of microsaccades is still an open question, but some points merit discussion. Are microsaccades connected to a visual optimization process that is driven by evolutionary selection? Tiny involuntary movements (i.e., those not triggered by conscious cognitive signals) may help prepare the eyes to perceive a new image that is not yet in focus, thus serving as an anticipatory preparation mechanism [237]. Such a mechanism would improve the organism's ability to switch between different spots, thus giving it an adaptive advantage. Note that visual adaptations benefit both predator (to rapidly detect prey against a background) and prey (to rapidly localize an approaching predator). Perhaps microsaccades are involuntary because voluntary movements are unnecessary and more costly neurophysiologically. Perhaps also, for identical reasons, they are micromovements rather than full saccades.

12.5 Learning, memory, and databases

Random searching also affects information searches. Information can be indexed and stored in highly organized and hierarchical information systems, e.g., dictionaries, encyclopedias, and catalogs, but the advantages of an index need to be balanced against the amount of energy and time required to create the index. Although one savant has memorized more than 3×10^4 digits of π, because of the cost of storing and indexing information most people choose not to memorize a long string of digits – or more generally, excessive information about a single event. Assuming that information has already been stored in some manner, the most efficient way to retrieve the information depends on how it is actually stored. The process of managing information can itself have random or stochastic aspects – notice where some people store receipts needed for filing income taxes!

Rhodes and Turvey [312] studied attempts by people to recall as many words as possible from a specific category, e.g., animal names. They found that retrievals occur sporadically over an extended period and decline as recall progresses but that short retrieval bursts occur even after many minutes of performing the task. They also found that for each participant in the experiment, the intervals between retrievals conformed to a power law tailed or Lévy distribution.

The more closely the exponent of the power law distribution of retrieval intervals approximated the optimal foraging value of 2, the more efficient the retrieval [312]. Rhodes and Turvey found that "at an abstract dynamical level, foraging for particular foods in one's niche and searching for particular words in one's memory must be similar processes if particular foods and particular words are randomly

and sparsely located in their respective spaces at sites that are not known a priori."
This raises the possibility that Lévy flight search dynamics, typically associated
with foraging, may also characterize retrieval from human semantic memory. Thus
a Lévy process could conceivably produce an adaptive advantage, this time related
to the cognitive power of *Homo sapiens*. Finally, note that $\mu = 2$ is a compromise
between short and long scales; therefore logically it would be the best choice if one
needed to remember day-to-day events and also not lose track of events further back
in time. Recall that $\mu = 2$ is a critical value that separates distributions with finite
mean flight sizes ($\mu > 2$) from those with diverging mean flight sizes ($\mu < 2$). In
fact, the mean flight size diverges logarithmically for $\mu = 2$ exactly.

12.6 Genetically modified crops and disease vectors

Epidemic propagation and pollen diffusion are complex topics beyond the scope of
this book. On the other hand, in the context of superdiffusion, what is noteworthy
is that disease vectors and pollen not only perform normal diffusion but may also
perform Lévy flights and superdiffusion.

Superdiffusion of genes from genetically modified (GM) crops is not inconceivable [338]. We are not focusing on the merits or drawbacks of GM crops, but
rather on possible unforeseen consequences. The most important point is that we
should not assume, *a priori*, that the diffusion of genes from GM crops will occur
at the rates predicted by the theory of normal diffusion. Instead, governments and
regulatory agencies should also take into consideration the superdiffusion scenario.

Similarly, models of disease that assume only normal diffusion can underes-
timate the motility of the disease vectors. We mention, in this context, citrus
variegated chlorosis (*amarelinho* in Portuguese, meaning "yellow"), caused by the
colonization of xylem by the bacterium *Xylella fastidiosa*. Models of the disease
now explicitly assume superdiffusion of sharpshooter vectors (insects) [229].

When infectious diseases in humans spread rapidly, unless evidence points to
normal diffusion, models with superdiffusion will give better predictions, especially
in worst-case scenarios. If we map the positions of the successive occurrences
of the H1N1 virus swine flu epidemic of 2009, we see that the epidemic was
superdiffusive. Part of the explanation for this is that humans – who travel in
superdiffusive movements – are themselves a vector for this virus.

13

Adaptational versus emergent superdiffusion

From the previous chapters, we see that (1) superdiffusion optimizes search efficiencies under specific (but common) circumstances and that (2) many animals move superdiffusively. Assuming these two facts, does it follow that there is a causal relation between them? Lévy strategies indeed optimize random searches, but does it necessarily follow that selective pressures systematically forced organism adaptation toward this optimal solution?

This is an important question because an adaptive pathway toward an optimal solution can prematurely stop at some suboptimal point that decreases the selection pressure on this particular feature to a level below the selective pressures on other issues [397]. Biology and physiology are replete with suboptimal solutions. The classic example is the structure of the human retina, which has blood vessels on the wrong side of the photosensitive layer [96]. Compromise solutions arise because adaptation (1) includes a stochastic component, (2) has to build on preexisting designs, and (3) occurs in a complex field where other pressures may be present and may possibly be stronger.

Dolphins, in the context of (mammalian) swimming adaptations, perform well, but how can we know whether or not their shape represents an optimal design? Some species of shark may have an even better hydrodynamic shape. Also, why did dolphins return to the ocean when selective pressures were pushing for improved terrestrial adaptation? The complex evolutionary history of real organisms contains many such contingent situations, such as changing selective pressures, genetic drift, low-number bottlenecks, and rare catastrophic events.

Returning to the original question, the encounter rate *per se* does not constitute a selective pressure; rather the pressure may involve improvements in the ability to find food or mates. From an evolutionary perspective, the issue is not so much whether Lévy flights are better than other strategies but whether a (possibly suboptimal) Lévy walk adaptation to improve stochastic searches could have been selected. In what follows, any discussion of Lévy walks as adaptive strategies should be interpreted in the light of these qualifications.

13.1 Are Lévy walks really adaptive?

Bartumeus *et al.* [19, 22] argued that Lévy walks confer adaptive advantages in relation to Brownian strategies and investigated the specific scenarios where this advantage may be important. In separate studies, Bartumeus *et al.* [21, 23] observed that the biological mechanisms optimizing the *chances of finding* targets are not necessarily identical to those responsible for improving the *detection* of targets; i.e., the selective pressures and the triggering stimuli in the two types of processes probably differ. Thus their combination may provide behavioral plasticity for adapting searches to diverse ecological scenarios.

One consequence of this decoupling is that random search strategies are not necessarily incompatible with short-range memory effects in time or space such as local scanning mechanisms or systematic rules of thumb. One suggestion is that scale-free *punctuations* in animal movement, such as stops and reorientations, could underpin the stochastic organization of the search at the landscape level [23]. For example, the benefits of random searching could be obtained by altering the frequency of random stops and adding correlations to reorientations in already existing search mechanisms based on local scanning. Bartumeus [17] maintains that scale-invariant reorientations can form the basis for stochastic search organization whenever perceptual capacities are substantially reduced (thereby increasing the relative importance of locomotion).

By studying a hybrid random walk with properties of both correlated random walks and Lévy walks, Bartumeus *et al.* [21] showed that the Lévy properties remain robust in behavioral mechanisms by providing short-range correlations in the motion, extending previous results [90, 394]. This separability between the scale-free properties, on one hand, and the short-range memory effects, on the other, could conceivably isolate the selective pressures acting on stochastic search mechanisms from other pressures. Hence specific stochastic behaviors could evolve more or less in isolation from other factors, thereby favoring their evolution.

This argument generalizes the Lévy flight hypothesis discussed in Chapter 10 because Lévy flights and walks are specific instances of scale-free behavior. If scale-free properties have adaptive value, then a genuinely scale-free adaptational property may exist underlying the strategies used to enhance biological encounters. Lévy flight foraging would be a special case of such scale-free behavior having adaptive value.

Fractional Brownian motion also has scale-free properties and has been used to try to describe the dynamics of movement of different organisms [147, 372], yet Lévy walks are more efficient than fractional Brownian motion in many random search scenarios [303]. One or both of these processes may constitute examples of an adaptation leading to scale-free strategies, where a suboptimal solution may be good enough.

13.2 Self-organization and emergence

The hypothesis that superdiffusive movement constitutes an adaptation contradicts the view that Lévy motion and scale invariance are epiphenomena, or emergent properties that arise solely via interaction with the environment. Although there are strong arguments that favor the adaptational viewpoint, we have not been able to rule out that the scale invariance observed in the movement of organisms may be a reflection of the scale invariance in the spatial structure of the environment (e.g., disorder). Similarly, it has been speculated [261] that adaptive search rules might have emerged from the nonlinearities in the interaction dynamics rather than through gradual evolution.

The hypothesis driving the emergent viewpoint is that the disorder in the environment (e.g., distribution of targets) determines the type of movement of the searchers. A helpful analogy would be the way the nuclear structure of an atom determines the main properties of Rutherford scattering.

When some of us [390] began in the early 1990s to study Lévy flights in animal movement (prior to the discovery in 1999 that Lévy flights can optimize random searches), our original intuition about why organisms might move superdiffusively was not the adaptational view but rather was strongly emergentist. We originally thought that the Lévy flights were the result of animals moving in structured (e.g., fractal) environments [390]. More recently, this view has been defended by Sims *et al.* [347] among others. For example, Benhamou [28] has presented an emergentist argument as an alternative to the Lévy flight foraging hypothesis, suggesting that even if Lévy walks are present, this does not mean that there is an underlying Lévy walk process. The Lévy behavior "may emerge from the way the animal interacted with the environment structure through more classical movement processes" [28].

Similarly, Reynolds [301], building on earlier work by Boyer *et al.* and others (see Section 13.3), has argued that chemotactically driven deterministic searches of random targets can result in power law scaling. This explanation for the prevalence of fat-tailed power law distributions in foraging data contradicts the view that organisms execute an innate optimal Lévy flight searching strategy. Reynolds argues that power laws can emerge from the tendency of chemotaxis to occasionally cause predators to miss the nearest target. This would not occur if targets were detected via a reliable cognitive map or if prey location were visually cued and always accurate. However, perfect information and detection are usually not possible because (1) the searching process is inherently stochastic and (2) natural selection of the targets (e.g., prey) can also occur, e.g., adaptations for escaping, evading, and hiding from the searchers. Reynolds [308] has also proposed that the power law scaling seen in turbulence may induce the observed power law scaling in the movement of organisms that move in fluids. Note, however, that some species prefer calm waters and avoid turbulence.

Most emergent theories and approaches assume a degree of heterogeneity or some other exogenous source for the leptokurtic behavior in the movement lengths. However, Gautestad and Mysterud [132] have argued plausibly that fractal, scale-free self-organization may arise due to spatial memory, even in homogeneous environments. In this context, we note that superdiffusion can arise not only via long-range spatial interactions (e.g., via ultralong jump lengths, as seen in Markovian Lévy flights) but also via long-range power law correlations in time (e.g., non-Markovian fractional Brownian motion). Indeed, long-range memory effects [88, 178, 180] can lead to anomalous diffusion and transport, as we discuss in Section 14.2.

13.3 Deterministic induction of Lévy behavior

We now discuss a specific case in which Lévy flights represent an emergent property. So far, we have discussed random walks with inherently stochastic dynamics and thus have assumed randomness *a priori*. A less widely known class of problems concerns deterministic walks [44, 45, 49, 66, 67, 124, 126, 182, 211, 314], which describe the movement of a *deterministic walker* in a certain environment and which may or may not have a random character. In this case, the rules of locomotion are always taken from a deterministic model rather than from a probability distribution [67], i.e., randomness is present only in the disordered media, never in the rules of motion. Note, however, that sometimes deterministic searches outperform random searches [170].

Deterministic walks in random environments occupy an intermediate position between purely random (generated by random trials) and purely deterministic (generated by deterministic dynamical systems, e.g., by maps) models of diffusion [67]. In this context, Boyer *et al.* [46] pioneered a novel approach to Lévy flight patterns by showing that deterministic walks can interact with complex environments in ways that allow a variety of memory effects, angular (i.e., orientational) correlations, and scale-free properties to emerge. Inspired by the earlier work of Boyer *et al.*, Santos *et al.* [322] studied the probability density function of step lengths

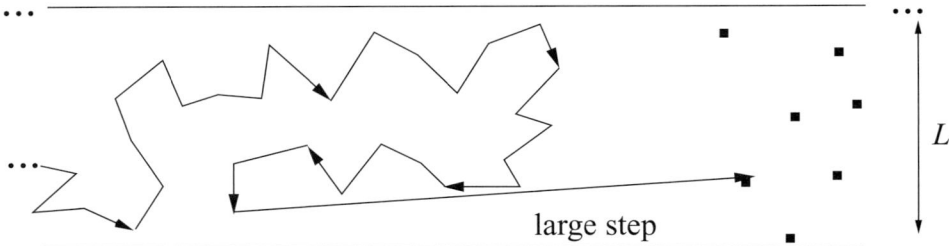

Figure 13.1 Deterministic walks in rectangular environments with randomly distributed targets. The go-to-the-nearest-site rule leads to power law distributions of move lengths for a specific critical aspect ratio for L. This surprising result is due to the fact that in the critical condition, the walker can accidentally backtrack and end up having to make an ultralong flight to return to the forward, more abundant region of target sites. Consequently, a non-Lévy process can lead to an emergent, scale-free behavior – via interaction with a disordered environment. This, in a nutshell, is the emergentist viewpoint.

taken by a deterministic walker that always travels to the nearest randomly distributed site in a closed rectangular geometry of fixed area $A = 1 = L_x \times L$. As expected, a characteristic scale of step lengths emerges for the one- (L very small) and two- ($L \sim L_x$) dimensional limits of the geometry.

Remarkably, however, a scale-invariant step length distribution is observed for specific aspect ratios of L and L_x in a striplike geometry. One might, in principle, expect the model to lead to a Poisson process, because the initial distribution of target sites (which are destroyed once visited) is random. Indeed, one finds that in both one- and two-dimensional limiting cases, the step length distribution has a finite variance. Nevertheless, for values of the order parameter L/ℓ_0 (with ℓ_0 being associated with the average minimum distance between two targets) in the interval $10 < L/\ell_0 < 30$, a nontrivial dynamical process with an arbitrarily large variance takes place, combining a large number of relatively small steps with a very few ultra long steps. Although the destruction of previously visited sites makes the walker move forward with a higher probability, there is also a finite fraction of large turning angles along the walk that allows the walker to backtrack. Eventually, the walker is forced to make an ultralong jump – the origin of the fat tail. Figure 13.1 illustrates this mechanism.

This result shows that a scale-invariant Lévy-like behavior can emerge as a consequence of interacting with the disordered media, even when the dynamics are driven by deterministic rules of movement – in this case, the go-to-nearest-site rule. In a biological context, emergence and adaptation need not preclude each other.

Nevertheless, in such deterministic models, the coupling or interaction between the searcher and the environment is extremely strong, and the walker always knows the location of the nearest target. This scenario is not realistic, but we are not aware of any studies on a more realistic model with weaker couplings.

13.4 Why the answer is crucial

We have argued in favor of two competing explanations for superdiffusive animal movement. This is important because if the adaptational view is eventually shown to be even partially correct, our understanding of the evolution of how animals move will require (possibly major) revision. It would mean that the statistical fluctuations in biological processes are themselves subject to selective pressure. Perhaps in the distant past, organisms could only move via normal diffusion governed by Gaussian statistics. Over evolutionary timescales, however, the biological apparatus for anomalous diffusion may have, due to selective pressures, slowly evolved. Hence the nature of the underlying statistics – e.g., Gaussian versus non-Gaussian – may itself be subject to natural selection.

It is difficult to imagine, for example, that visual microsaccades had non-Gaussian characteristics from the beginning (see Chapter 12). The neurophysiological mechanisms necessary for generating truncated power law fluctuations in microsaccades must have taken a long time to evolve. In the remote past, they most likely obeyed Gaussian statistics, and adaptations for power law behavior must have, at some point, appeared through natural selection.

Moreover, the stochastic effects and relevant adaptive drives could percolate upward in the organization and dynamics of ecosystems, e.g., to the population level (possibly even modulating biodiversity). We also cannot rule out the possibility that they affect the course of events on long evolutionary timescales.

Lévy walks could be a factor in extinctions – even mass extinctions. This possibility has been tackled in terms of autonomous random walkers whose sole source of energy are targets that are themselves diffusing random walkers [115]. Using numerical simulations and a simple mean field analytical model to analyze how the energy accumulated by the searcher varies with the target density, it was found that superdiffusion of either searcher or target confers substantial energetic advantages to the searcher. We saw in Chapter 10 that superdiffusion does not play a major role when target densities are high, but it confers a vital advantage in the limit of low densities at the edge of extinction. Specifically, diffusive searchers rapidly die, but superdiffusive searchers can survive for long periods without becoming extinct. Could such effects be a factor in mass extinctions?

The general consensus is that dinosaurs became extinct 65 million years ago because of a single, massive asteroid impact or multiple nearly simultaneous impacts, but why did some organisms survive this event and not others? Except for those that would become the ancestors of birds, all dinosaurs went extinct, but not all other reptiles or mammals did. Why were the results of the impact not the opposite – the dinosaurs surviving and the mammals dying? An impact could have been the starting point (i.e., the trigger) or proximal cause for the death of the dinosaurs, but what was the long-term, ultimate cause? It is possible that the relative diffusiveness of the various species played a role in determining which animals survived and which did not.

Chapter 6 contains examples of organisms that appear to increase diffusivity when the target density decreases. Under harsh environmental conditions, species presenting such adaptive behavior may have a survival or mating advantage over those that do not [397]. The type of movement may determine which species perish and which survive under extreme conditions. It is plausible that this selection pressure could have been a factor in the dinosaur and other extinctions, e.g., the greatest mass extinction on Earth, the Permian-Triassic event. In fact, Faustino *et al.* [115] studied the effect of superdiffusion on the death rates of random walkers that eat other random walkers (i.e., prey) to obtain metabolic energy. The study found that whereas ballistic and superdiffusive behaviors have a negligible effect when prey density is high, they become crucial when prey density is low and the forager is facing possible extinction.

Recall that for destructive foraging, ballistic interactions are intrinsically more efficient than diffusive interactions. Organisms tend to maximize encounter rates with their prey and minimize them with their predators. Therefore, the theory predicts that organisms will display motility with correlation run lengths that are long compared to the reaction distances to their prey but short compared to the reaction distances to their predators [388]. Visser and Kiorboe [388] analyzed motility data for planktonic organisms (ranging from bacteria to copepods) that support this prediction. This agreement between theory and experiment supports the adaptational argument. What is even more convincing is that the chemotaxis networks of some bacteria have noise generators that appear to have been selected [189]. If they are there by design, they must be adaptations.

Finally, it is possible that the adaptational and emergent viewpoints are not mutually exclusive [397], but if future research establishes superdiffusive movement as an evolutionary adaptation in some species, it would represent a stunning (if not entirely unexpected) development in movement ecology, raising new questions and adding kinematic, transport, and diffusive dimensions to existing problems (e.g., the colonization of Madagascar; see Section 7.5).

14

Perspectives and open problems

14.1 The flavor of foraging research

The physics of foraging is exciting because so little is known and so many questions remain. Biological foraging and random searching are relatively new fields, and considerable effort is still being made to establish theoretical foundations and reliable and general methods of data collection and analysis. Many challenges are still to be overcome, most of them related to technical issues and interpretation of findings.

In this final chapter, we put the major open problems into perspective. One reason for the skepticism about anomalous diffusion and Lévy flights is the lack of obvious biological mechanisms for generating superdiffusive random walks. We will also discuss the issue of free will and the existence and uniqueness of globally optimum strategies. We begin with two problems currently being studied by researchers.

Foraging on the edge of extinction

In mass extinctions and smaller-scale extinctions, the density of organisms becomes zero (if extinction is total) or very low (if recovery eventually takes place). We saw in Chapter 13 that as the density of targets lowers, the importance of superdiffusion increases. How does organism movement change during extinction events? Is there any change in the selection pressure on how organisms move? Such questions remain relatively unexplored at the present time.

Lévy searches on small-world networks

We saw in Chapter 10 that Lévy motion can confer advantages to search processes not only in Euclidean spaces but also in discrete analogues (large-world networks), but what happens in the environment of small-world networks?

A single random walk step along a link connecting distant nodes can cover a great distance. Brownian searches on small-world networks behave in the same manner as Lévy searches on large-world networks. The advantages conferred by Lévy flights are probably diminished in small-world networks – the main factor being the structure of the links and the power law exponent for the link distribution – but this hypothesis remains to be tested.

Variations on a theme

The preceding two examples show how we approach the study of biological movement from the perspective of reaction-diffusion processes. We ask what happens when some aspect of an existing problem whose solution is known is changed.

Note that we do not ask which of several models of movement is correct, or how organisms move (or should move) *per se*, but whether there is a change in how they move – whatever it may be – when the organism density plunges. Note that we pay careful attention to the role of the underlying embedding space. Recall that the non-Euclidean Internet differs greatly from a regular lattice, and thus we expect that the optimization process of random searches on the Internet will differ qualitatively from the process on regular lattices.

This approach helps to establish the parameters of the field and advance our understanding through incremental progress. For example, by modifying the different parameters of a foraging scenario (e.g., type and density of targets, geometry, availability of information, existence and nature of correlations), we can identify the conditions and strategies that optimize the search. Apart from this approach and what it can accomplish, there are other problems that stand as central questions in their own right. We explore some of these major issues in the following sections.

14.2 Biological mechanisms underlying superdiffusion

One of the most interesting areas of research concerns how biological mechanisms modulate movement. The discovery of the actual mechanisms that generate superdiffusive movement will be relevant in the ongoing debate between the emergentist and the adaptationist viewpoints and will go a long way toward reducing resistance to ideas of superdiffusion and Lévy flights. We review here some recent developments.

Environmentally induced superdiffusion

Superdiffusion does not require a specific mechanism for its generation. Other mechanisms or adaptations with other purposes can generate superdiffusion as a side effect.

Conspecific avoidance

One biological mechanism that may generate Lévy flight patterns is avoidance of locations previously visited by individuals of the same species. Guy *et al.* [145] argued that this *conspecific avoidance* behavior may generate optimal Lévy flight (scale-free) search patterns when prey is sparse and randomly distributed.

Conspecific avoidance reduces oversampling and makes this hypothesis plausible. Reynolds [300] has shown how optimal Lévy flight patterns can emerge from conspecific avoidance. Guy *et al.* [145] studied the movements of a starved carabid beetle and found that when it walked on objects that other beetles had previously visited and also on unvisited control objects, there was experimental evidence of conspecific avoidance, i.e., the beetle preferred the unvisited control objects [145]. This finding supports the hypothesis that the decrease in oversampling caused by conspecific avoidance benefits starving organisms searching for food.

Chemotactically driven superdiffusion

Although the avoidance of previously visited sites (i.e., oversampling) constitutes a plausible biological mechanism for generating Lévy walks and superdiffusion, we would not expect mechanisms that do not reduce oversampling to generate Lévy walks. Reynolds [301] studied chemotactically generated Lévy walks and reported that Lévy flight patterns with an inverse square power law scaling emerge naturally in deterministic walks performed by predators that use chemotaxis to locate randomly and sparsely distributed targets. The proposed mechanism underlying the emergence of scale-invariant distributions of displacements is the propensity of chemotaxis to miss the nearest target.

Internally generated superdiffusion

Conspecific avoidance and chemotaxis are both presumed to generate superdiffusion via interactions with the environment, e.g., other organisms, but superdiffusion can also arise directly from internal dynamics, i.e., endogenously.

Feedback with long-range memory

Anomalous diffusion can arise from delay differential equations [71, 88, 177]. Such delays in stochastic processes appear naturally in nervous systems due to the finite velocity of signals propagating along nerve pathways. Together with positive or negative feedback, such delays can result in anomalous diffusion, including superdiffusion [88].

Surprisingly, superdiffusion can arise not only via long-range memory leading to directional persistence but also via the loss of such memory. Cressoni *et al.* [88] have shown that non-Markovian walks can suddenly switch to (log-periodic) superdiffusive behavior when there is a large loss of memory of the recent past. Kenkre [177] coined the term *Alzheimer walk* for this behavior. It has been speculated that this onset of repetitive behaviors may bear a causal relation to a loss of memory of the recent but not distant past.

A repetitive pattern in movement can lead to superdiffusion. Whether similar mechanisms actually occur in biological systems that modulate movement is an unanswered question. Also unknown is their adaptive value (if any).

Decision-based queuing

Vazquez *et al.* [384] studied such human activities as e-mail and letter-based communications, Web browsing, library visits, and stock trading. They found that the timing of individual human actions follows non-Poisson statistics and is characterized by bursts of rapidly occurring events separated by long periods of inactivity. They showed that these bursts in activity are due to a decision-based queuing process. Prioritization allows most tasks to be rapidly executed, but some tasks are postponed for a very long time.

Bartumeus *et al.* [21] explored some possible biological mechanisms capable of generating Lévy walks and concluded that Lévy motion is sustained when processes for punctuating the movement with sudden reorientations are present. Because bursts of rapidly occurring reorientation events separated by long periods with ballistic motion can lead to superdiffusion, we ask whether organisms use prioritization processes to generate superdiffusive motion.

Neurophysiological mechanisms

Assuming these exist, the question is not whether neurophysiological processes drive superdiffusive movement patterns but rather how they succeed in generating superdiffusion or persistence (see [61]). One example of a biological mechanism capable of generating Lévy-distributed random variables comes from a study by Cabrera and Milton [69] of a stick being balanced on a fingertip [70]. High-speed motion analysis in 3D of human subjects showed a Lévy flight of fingertip motion, with $\mu \approx 2$, and showed that the truncation cutoff for the Lévy flights became larger as skill increased. This increase in the cutoff with increasing skill has deep implications for balance control problems, ranging from falling patterns of the elderly to the design of two-legged robots.

The only realistic explanation for their finding is that the human nervous system interacting with the environment in specific ways can generate Lévy flights. The basic mechanism appears to involve brief intervals of corrective movements alternating with longer intervals of more free movement. If the act of balancing a stick can generate a Lévy flight, then we can infer that a moving organism can also execute Lévy flights, the essential ingredients being the alternation of orientationally corrective events that punctuate otherwise free (perhaps ballistic) movement.

Martin *et al.* [225] showed that the previously reported scale-free power law behaviors of fruit flies (*Drosophila melanogaster*) disappear if the synapses established by neurons in the ellipsoid body are specifically blocked. They concluded that the ellipsoid body (i.e., the so-called fly motor cortex) regulates the fractal temporal pattern in fruit flies.

The scale-free microsaccades in the visual system discussed in Chapter 12 probably have a similar neurophysiological basis. Indeed, it is difficult to imagine any other explanation (e.g., an emergentist explanation). Similarly, Harnos *et al.* [148] found that the time spent by a young female nesting pig (gilt) in a given form of activity has a power law tailed distribution and argued that the source of this complex behavior is the neural system, forced by a hormonal stimulus associated with nesting instincts.

As mentioned earlier, Bartumeus has argued that reorientation mechanisms may be the missing ingredient needed to connect the levels of description of trajectories with internal neurophysiology. Note that this idea is empirically falsifiable. Because reorientations are detectable along a trajectory, such hypotheses can be tested.

We still do not know which specific neurophysiological mechanism is responsible for the generation of Lévy walks and other superdiffusive motion. Reynolds [303] noted that inverse square power law distributions of spontaneous neuron firing signals may provide the timing signals necessary for the execution of Lévy searches. Levina *et al.* [208] showed analytically and numerically how self-organized critical [14] behavior could arise in neuronal systems. Such scale-free signals, which could form the basis for the patterns of punctuating reorientations needed to generate Lévy walks, have been observed in *in vitro* studies [25, 235, 334]. Despite such circumstantial evidence, there is no direct evidence yet linking *in vitro* behavior with Lévy walks of freely roaming organisms. The discovery of such a direct link would have a strong impact on the field of movement ecology.

14.3 Determinism, randomness, and free will

Although free will is a nebulous concept, our current legal systems and religious traditions make frequent reference to it. Because society is heavily invested in the idea and it is hard to study objectively, perhaps the study of free will (or its analogue) in animals can help us better understand free will in ourselves. A foraging animal *chooses* a path from a pool of possible paths. Even the simplest organisms make regular choices. The degree of autonomy seen in such decision making bears a striking resemblance to what we usually think of as free will in humans.

The physical basis of free will remains an enduring mystery, even though all organisms are subject to physical laws. Classical (i.e., Newtonian) mechanics for a closed system is deterministic. Quantum mechanics supplanted classical mechanics and differs from it in important ways, perhaps the most important being the introduction of intrinsically probabilistic (i.e., stochastic) aspects. Moreover, quantum mechanical states can decohere [418]. Some interpretations of quantum mechanics, e.g., by Bohm, are deterministic, but theoretical and experimental results related to Bell's inequality strongly support the conventional probabilistic interpretation. One could thus try to argue that quantum mechanics allows free will. Still, it is difficult to see how randomness by itself can lead to free will. Can an open quantum mechanical system with long-range memory process or "metabolize" randomness such that its behavior is neither purely deterministic nor purely random but instead exhibits a deterministic process subject to frequent stochastic perturbations? This kind of behavior would avoid the concept of destiny. The frequent stochastic perturbations could arise due to interactions with the environment – and lead to decoherence and quantum mechanical measurements. Such a system would have endogenous and possibly non-Markovian dynamics. The question is whether such endogenous non-Markovian behavior is what we perceive as free will. A more dramatic possibility is that, in the future, quantum mechanics will be superseded by a more complete physical theory, a belief expressed by both Dirac and Einstein. In this context, Penrose [277] has argued that quantum mechanics and general relativity cannot be reconciled without a new fundamental theory of quantum gravity, which might lead to new insights into nonalgorithmic aspects of consciousness and free will.

An even more dramatic view is that perhaps what we experience as free will is ultimately an illusion (e.g., an epiphenomenon). Free will is constantly under attack, most recently by advances in neuroscience; e.g., freely willed decisions seem to enter consciousness after the fact. Libet's [210] seminal experiments strongly suggest that volitional actions are mistakenly interpreted as having conscious motivation, making the subjective experience of free will a consequence of mistaken retrospection.

However, a recent essay published in *Nature* by Heisenberg [151] defends free will by invoking animal behavior. Inspired by Kant, Heisenberg's operationally convenient definition states that free will is an action that is self-initiated or self-generated. Animals and organisms appear to undertake self-initiated action that is not a response determined by the environment. Such self-initiated action appears random, i.e., it is different each time an experiment is repeated. In contrast, a determined response is the same every time, hence, predictable. Heisenberg's working definition of free will neither conflicts with physical theories nor ignores consciousness. Note that self-initiation is not necessarily an irreducible property and that we must not forget the distinctions between proximate and ultimate causes of behavior. If quantum mechanics is correct, then the ultimate causes of self-initiated action must be traceable to a mix of prior events and randomness. This last statement leaves open the possibility that free will is actually conditioned and not really free.

The empirical study of biological movement similarly raises further questions about the concept of free will. Why should animals having a degree of free will obey any physical law whatsoever? The issue is analogous to human behavior in financial markets. We assume that each buyer and seller has free will, yet financial returns and financial volatilities can be described statistically [222].

Heisenberg's operational definition of free will as self-initiated action does away with questions concerning the subjective experience of free will, i.e., of our own awareness of steering events according to our preferences. Human self-initiated action without consciousness is not what we normally call free will. Sleepwalkers may carry out self-initiated action, but the common view is that free will is not present during sleepwalking. In 1987, a man committed a murder while sleepwalking. The event happened in Canada and involved Kenneth Parks, who drove more than 20 km and stabbed his mother-in-law to death. He had had good relations with his mother-in-law, claimed at the trial that he had been sleepwalking and, because of the EEG evidence and lack of any motive, was acquitted. To what extent do sleepwalkers have free will? Can they be held morally responsible for their actions if they are not awake? Studying how sleepwalkers move may reveal intriguing aspects of unconscious self-initiated movement but may do little to help us better understand our conscious experience of free will.

In summary, the study of how animals move autonomously may help to shed light on the seemingly intractable problem of free will. On one hand, a quantitative (possibly statistical) description of free will as self-initiated action is well within the reach of current scientific methodology. On the other hand, it still remains a mystery how objective physical laws can be reconciled with our subjective experience of free will.

14.4 Globally optimum random searches

We saw in Chapter 9 that the random walk propagator contains information relevant to search optimization. The important ingredient for optimal nondestructive searches is a large Hurst exponent, $H > 1/2$. Lévy walks with fat-tailed propagators are one such example. For all practical purposes, the precise mechanism that generates the fat tails in the propagator does not matter in the context of optimization. What matters is where the random walker diffuses, not the manner by which the walker gets there. However, in addition to Lévy processes, other stochastic processes, e.g., low-dimensional chaos and noise-induced chaos, may generate scaling [127]. Fractional Brownian motion, for instance, can also be superdiffusive but differs from Lévy walks and flights.

These distinctions point to a deeper question. Are there strategies that outperform the best ($\mu = 2$) Lévy walks? One of the most important open theoretical problems concerns the globally optimum choice of random walk processes for nondestructive searches.

14.5 Final remarks

Empirical data: State of the art

The increasing availability, quality, and quantity of publicly accessible data on animal movement will allow many open questions to be properly addressed and answered in the near future. A considerable number of advances are already taking place in theoretical movement ecology, and we expect many new and perhaps unexpected advances to take place in the next couple decades.

The most recent empirical study, by Humphries *et al.* [163], involves the rigorous statistical analysis of more than 10^7 data points covering more than a dozen species. Their data set is an order of magnitude larger than the previous one [347]. They find strong evidence of Lévy walks. However, as predicted theoretically, they are not universal.

Epilogue: The main challenges for the future

In order to set realistic research goals for the near future, we list what we consider to be the three most important open problems:

1. To what extent and under what circumstances do organisms move superdiffusively?
2. Is such superdiffusive movement an adaptation or an epiphenomenon – an emergent property?
3. What is the globally optimum search strategy for locating scarce, random, and uniformly distributed targets?

At present, we are close to answering conclusively only the first question. The other questions remain open. Our sincere hope is that this book will stimulate wide discussion and lead to new investigations of these challenging problems.

Appendix A

Data analysis

A.1 A criterion for inferring superdiffusion

Here we review the criterion for establishing superdiffusion reported in [396]. We argue that this is a necessary but not a sufficient condition.

Consider two-dimensional correlated random walk (CRW) models, in which persistence (or directional memory) is controlled by the probability distribution of the relative turning angles. We define the turning angle θ_j as the difference between the angles of successive step vectors $\mathbf{r}_{j+1}$ and $\mathbf{r}_j$. If the turning angles are uniformly and independently distributed, then the path follows the usual uncorrelated Brownian random walk. However, if the turning angles are nonuniformly distributed, then even for independently and identically distributed turning angles, the resulting random walk step vectors $\mathbf{r}_j$ may be autocorrelated. In the extreme case, if the probability density function (pdf) for the turning angles is given by a Dirac δ-function at $\theta = 0$, the resulting walk is a straight line; i.e., the motion is ballistic.

However, generally, CRW models cannot have scale invariance. Instead, CRWs possess a characteristic scale (or time τ) associated with the exponentially decaying correlations always present in Markov processes. To obtain τ, we define an adimensional two-point correlation function in terms of the random walk step vectors $\mathbf{r}_j$ as $C(|j - i|) \equiv \langle \mathbf{r}_j \cdot \mathbf{r}_i \rangle / \langle r_j r_i \rangle$, where i, j are integer indices representing time, assuming constant speed. The notation $\langle \cdot \rangle$ denotes averaging (either averaging along the walk or ensemble averaging, depending on the context).

We assume that the CRW has mutually independent and identically distributed step lengths r_j (of finite variance) and turning angles θ_j, so $C(1) = \langle \mathbf{r}_j \cdot \mathbf{r}_{j-1} \rangle / \langle r_j r_{j-1} \rangle$ reads

$$C(1) = \langle \cos[\theta] \rangle = \int_{-\pi}^{+\pi} d\theta \, \cos[\theta] \, f_{\mathrm{w}}(\theta), \qquad (A.1)$$

with f_{w} denoting the circular "wrapped" (pdf) of relative turning angles. For circular statistics, the mean resultant length ρ and mean direction $\overline{\theta}$ are related to the first circular characteristic function by $\phi_1 \equiv \langle \exp[i\theta] \rangle = \rho \exp[i\overline{\theta}]$. Unless $\overline{\theta} = 0$, the CRW will contain loops that prevent persistence. Therefore we can restrict our attention to $\overline{\theta} = 0$ so that $\rho = C(1)$. Note that $\rho = 1$ for a Dirac δ distribution and $\rho = 0$ for a uniform distribution. The Markovian character of CRW implies

$$C(t/t_0) \sim [C(1)]^{t/t_0} = \exp[(t/t_0) \ln[\langle \cos[\theta] \rangle]], \qquad (A.2)$$

where t_0 is the typical time of one step. We thus have

$$\tau = -1/\ln[\langle\cos[\theta]\rangle] \qquad (A.3)$$

as the adimensional correlation time (or length) measured in step units. In other words, τ is the mean life of the correlations. The validity of this expression extends to all one-step Markovian CRW models. For any scale a few orders of magnitude larger than τ, the CRW appears Brownian because the model cannot keep the orientations correlated at such relatively large scales [23, 396].

The natural question, then, is whether a data set supports genuine superdiffusion (e.g., a Lévy walk) as opposed to a CRW, which only appears superdiffusive at sufficiently small scales. Our results suggest a natural criterion for determining when a data set contains enough information to answer this question. Because a CRW only converges to Brownian motion on scales much larger than τ, any data set spanning a period Δ not much larger than the correlation time does not contain sufficient information to make such a distinction with any level of statistical significance. We can estimate

$$\tau_{\mathrm{meas}} \equiv -1/\ln[\langle\cos[\theta]\rangle_{\mathrm{meas}}], \qquad (A.4)$$

in which the expectation $\langle\cos[\theta]\rangle_{\mathrm{meas}}$ denotes the experimentally measured value of the first cosine moment. Because we can infer τ_{meas} from a relatively small stretch of data, in practice, we can calculate an upper bound for τ. Hence the following necessary but not sufficient condition for establishing superdiffusive behavior has been proposed:

$$\Delta \gg \tau_{\mathrm{meas}} = -\frac{1}{\ln[\langle\cos[\theta]\rangle_{\mathrm{meas}}]}. \qquad (A.5)$$

Specifically, if a given finite data set corresponds to a timescale Δ that is two orders of magnitude larger than the value of τ_{meas}, then we can, in principle, distinguish CRWs from true superdiffusive walks. Indeed, a CRW process at this scale would contain approximately 10^2 or more effectively independent stretches, thereby providing adequate statistics [23].

It is easy to see why we need approximately two orders of magnitude. The value of τ is the mean life of the correlations, so at timescales one order of magnitude larger than τ, a CRW will appear uncorrelated. Because mean squared displacements are calculated by averaging along the walk rather than by taking ensemble averages, we need an order of magnitude of at least 10 independent segments of the walk to compute the averages, each segment having a length of at least 10τ. So at scales of 100τ, every CRW will begin to show diffusive behavior (due to the central limit theorem). In contrast, a genuinely superdiffusive walk shows directional persistence even at such a long timescale.

Assuming constant velocities, a superdiffusive walk will have a correlation time that diverges, but the estimated value τ_{meas} via $\langle\cos[\theta]\rangle_{\mathrm{meas}}$ can never diverge unless the turning angle distribution is a Dirac δ function. Recall that, for walks with finite velocity, the central limit theorem will apply unless random walk steps separated arbitrarily in time remain correlated. For a random walk with finite step sizes to possess the property of superdiffusion, the random walk steps must possess algebraically decaying correlations such that the correlation time diverges. If the correlation time does not diverge, then the random walk will appear Brownian at timescales very much larger than the correlation time. However, for any turning angle distribution (excepting the Dirac δ function), the correlation function of a CRW decays exponentially; hence the correlation time of a CRW cannot diverge.

If a given data set does not span a long period, only indirect methods of inferring genuine superdiffusion can be employed such as tests of self-affinity or direct estimation

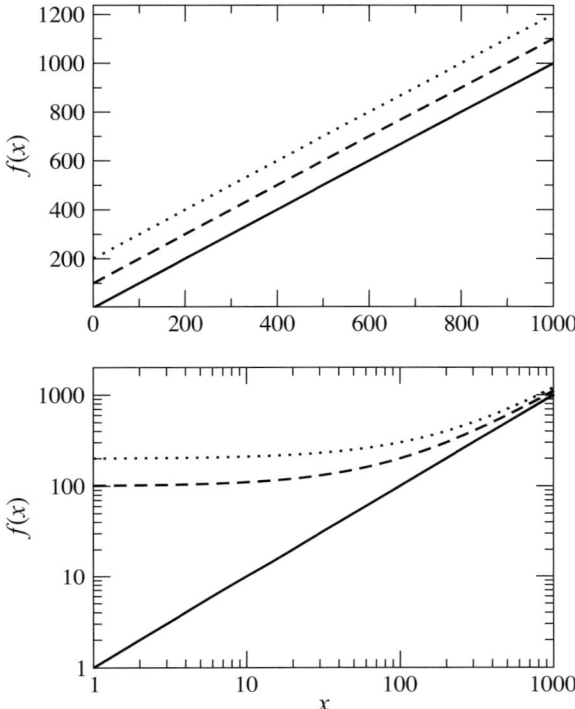

Figure A.1 The functions $f(x) = x, x + 100$, and $x + 200$ shown on linear and double log plots. A power law plus constant does not appear linear in double-log scale. Such artifacts of double-log plots must be taken into account carefully when analyzing data.

of the correlation function to check for long-range power law decay of the turning angle memory retention. On the other hand, if a given data set satisfies the criterion, then a test of superdiffusion on scales larger than τ_{meas} can eliminate possible false positives for shorter scales that arise due to Markovian turning angle persistence.

A.2 Log-log plots and surrounding controversies

How reliable is the use of double-log plots – say, of $\ln f(x)$ versus $\ln x$ – to establish power law scaling $f(x) \sim x^\beta$? If care is not taken, several difficulties may arise when dealing with such plots. A detailed discussion of these issues is beyond the scope of this book, but in what follows, we discuss the main aspects of data analysis when studying animal movement.

Pure power law functions of the form

$$f(x) = \text{constant} \cdot x^\beta \tag{A.6}$$

become linear on a double-log scale, but analytical problems arise when the function is not a perfect power law (e.g., gradually truncated power laws) because there will be curvature in the corresponding double-log plot. For example, as shown in Figure A.1, a constant added to a power law shows up as a nonlinear distortion in the double-log plot:

$$\log[\text{constant} + x^{\beta}] \neq \text{another constant} + \beta \log x.$$

Similarly, all smooth functions will appear linear on a log-log plot if the range of scaling is small. Therefore the method could lead to misinterpretation.

Note that an underlying power law may be missed because of nonlinear scaling in the plots – and conversely, a power law may be inferred where there is none. These difficulties become especially problematic when there are fewer than three decades of scaling in both axes. Many physical systems, e.g., earthquakes and second-order phase transitions, exhibit scaling over many decades, so the problem does not arise. In contrast, data for animal movement rarely contain more than three decades of scaling. Therefore data analysis must be undertaken with extra care.

As an alternative to log-log plots, statistical inference has been proposed for testing for power law scaling in movement data [105]. The underlying premise is that graphical methods are less precise than statistical inference. However, there is no consensus yet that statistical inference is any better than using log-log plots. Indeed, power laws are always linear on log-log plots. There is no controversy over using log-log plots to study earthquakes or second-order phase transitions with many orders of magnitude of scaling. The problem may not be the method but the quality and quantity of the data. A detailed discussion of plotting methods can be found in a comparative study by Sims *et al.* [346]. They discuss such methods as logarithmic binning, a technique common among physicists but less common in other fields.

For such reasons, we recommend the use of both graphical and inference-based methods. Graphical methods, including log-log plots, give readers an immediate qualitative picture. However, they are often too crude for precisely extracting quantitative information, especially in small data sets (e.g., the values of scaling exponents). As an illustrative example of the strengths and weaknesses of log-log plots, consider Kleiber's law, which states that the metabolic rate of animals scales approximately as the body mass to a fractional power. The power law relationship was immediately understood by anybody seeing the log-log plot contained in Kleiber's original work [186]. On the other hand, the value of the scaling exponent continues to be debated even today, more than half a century later. Indeed, it is not easy to use standard approaches, such as linear regression, to fit power laws. Because it is universally known that a power law can be linearized by taking logarithms, some mathematical software applications with fitting algorithms attempt linear regression on the logarithms of the relevant variables. Standard linear regression minimizes the square errors, but this is not what happens if logarithms have been taken prior to linear regression. Unless unequal weights are explicitly taken into account during the least squares regression, the best fit will not, in fact, minimize the square error. Indeed, power laws pose peculiar and particular challenges to data analysis and interpretation. In this larger context, statistical inference methods are a welcome additional tool. For instance, maximum likelihood estimation seems to give the best values for scaling exponents (see Section A.3). Given the somewhat complementary strengths (and weaknesses) of graphical methods and inference methods, the most comprehensive studies [163, 347] of animal movement make wide use of both graphical methods and statistical inference.

A.3 Maximum likelihood estimation

When checking a model against given data, a useful tool is maximum likelihood estimation (MLE), which provides the values of the model parameters that optimize agreement with the data, given a particular data set. MLE has found many successful applications in statistical inference, but here we discuss only the application of statistical inference to systems having power law scaling.

Statistical inference in general, and MLE in particular, are valuable tools for extracting information from data and for comparing different models. For example, a comparison of different methods for estimating power law exponents has found that MLE gives the best estimates [409]. Nevertheless, there are known issues with MLE that must be taken into account if we are to minimize the frequency of misleading results. We summarize some of these issues in this section.

For data sets of small size, MLE is unreliable because of bias effects. This problem becomes compounded when using MLE to study Lévy flights and power laws in general. The additional complication arises because of the meaning of *small size*.

Finite size effects related to power laws are general and not limited to MLE. Consider, for example, an uncorrelated random walk consisting of N steps of unit length. We see that the 10-fold convolution of a uniform distribution already looks Gaussian, i.e., its envelope is approximately a Gaussian (i.e., normal) distribution. The width of the Gaussian will be of the order of magnitude of $\sqrt{N}$ steps. In contrast, the width of an α-stable Lévy distribution will be $\sim \sqrt[\alpha]{N}$. The fact that the Lévy distribution grows faster with N means that the events are distributed over a wider domain, so the statistics become thinned out. Specifically, a Lévy walk having traversed a distance of N unit lengths can contain only a single Lévy jump of size N, 2 of size $N/2$, and n of size N/n. The tails of the Lévy distribution are thus disproportionately affected. In contrast, for a Gaussian distribution ($\alpha = 2$), the tails are exponential, so one can effectively ignore the data, say, 10 standard deviations from the mean. However, for power law distributions, we cannot ignore the tails, so the meaning of *small size* depends on the distribution.

Note that any statistical inference method that gives equal weight to large and small events will have a built-in bias against power laws. Consider, for example, a probability density function for a random variable that appears Gaussian within 3 standard deviations but that has a heavy power law tail that extends up to 10 standard deviations. There are many more data points in the Gaussian-like region, so if this effect is not taken into account, the power law tail may be missed. Equivalently, one can say that large but rare events play a nonnegligible role whenever scale-invariant power laws are involved. We can understand this by using an analogy: although there are 1000 times more earthquakes with magnitude 4 than with magnitude 7, it would be a mistake to assume that the magnitude 4 earthquakes are more important than those with magnitude 7.

Finally, we return to the issue of truncated power laws. In nature, power laws appear truncated because singularities and infinities are not observable, even in physical systems (see Chapter 6). There is always a cutoff. Neglect of truncation effects, combined with the other problems mentioned earlier, leads to incorrect or misleading results. Blind or careless use of MLE in systems governed by power laws or truncated power laws may thus lead to incorrect inferences.

Appendix B

Lévy walkers inside absorbing boundaries

We review here the formal treatment of the time spent by a Lévy walker inside a container with absorbing boundaries, on which the approximate scaling arguments given in Chapter 10 have their basis. We follow the approach developed by Buldyrev *et al.* [62, 63].

Consider a Lévy flyer that starts at point x_0 in a one-dimensional search space with boundaries (absorbing target sites) at $x = 0$ and $x = \lambda$. In Chapter 10, the efficiency of the search was written in terms of the (inverse of) the average distance traversed between two successive targets. Here the cases of nondestructive and destructive searches correspond, respectively, to $x_0 = \ell_0$ (having found the site at the origin, the searcher restarts at the minimum possible distance from it, if we set $r_v = \ell_0$) and to $x_0 = \lambda/2$ (since the last visited site is destroyed, the searcher restarts, on average, at the middle of the interval). We now calculate $\langle L \rangle$ and the mean number of flights between two successive target sites, $N \equiv \langle n \rangle$.

The probability density $P_n(x)$ of finding the Lévy searcher at point x after n flights can be expressed recursively in terms of initial probability density, $P_0(x) = \delta(x - x_0)$, as

$$P_n(x) = [\mathcal{L}_\mu^n P_0](x), \tag{B.1}$$

where the integral operator $\mathcal{L}_\mu$ acts on a function f as

$$[\mathcal{L}_\mu f](y) = \frac{(\mu - 1)\ell_0^{\mu-1}}{2} \int_0^\lambda \frac{f(x)\theta(|x - y| - \ell_0)dx}{|x - y|^\mu}. \tag{B.2}$$

We have used here the step function $\theta(x) = 1$ for $x > 0$, and 0 otherwise. We will neglect the effects of absorption when $0 < x < r_v$ or $\lambda - r_v < x < \lambda$ and instead assume that absorption happens only at the extremities. By considering the probability that the flyer is absorbed (a target is found, either at $x = 0$ or $x = \lambda$) exactly on the nth flight,

$$\tilde{P}_n = \int_0^\lambda [(\mathcal{L}_\mu^{n-1} - \mathcal{L}_\mu^n)P_0](x)dx, \tag{B.3}$$

and integrating along with the Dirac δ function of P_0, we find a closed-form expression for the average number of flights before absorption:

$$N \equiv \langle n \rangle = \sum_{n=1}^\infty \tilde{P}_n n = [(\mathcal{L}_\mu - \mathcal{I})^{-1}h](x_0), \tag{B.4}$$

where $\mathcal{I}$ is the unity operator and $h(x) = -1$. Equation (B.4) can be rewritten in terms of a Fredholm integral equation of the second kind [351] with kernel $p(x_0 - x_1)$:

$$Q(x_0) = \langle q_0(x_0) \rangle + \int_0^\lambda Q(x_1) p(x_0 - x_1) dx_1. \tag{B.5}$$

The kernel p is actually the probability density function for the flight length distribution and $Q(x_0) = \langle \sum_{i=1}^\infty q_i \rangle$, $q_i = q(x_{i-1}, x_i)$ is a function of the starting point x_{i-1} and ending point x_i of the ith flight, $\langle \; \rangle$ denotes the average over all possible processes starting at x_0, and $\langle q_0(x_0) \rangle \equiv \int_{-\infty}^\infty p(x_1 - x_0) q(x_0, x_1) dx_1$, noting also that if x_1 lies outside the interval $[0, \lambda]$, the searcher is absorbed by one of the boundaries, and the value of $q(x_0, x_1)$ should be defined according to its physical meaning in the search process. In particular, the preceding calculation corresponds to the assignments $Q(x) = \langle n(x) \rangle$ and $\langle q_0(x) \rangle = -h(x) = 1$. Moreover, a similar procedure using $Q(x) = \langle L(x) \rangle$ and $\langle q_0(x) \rangle = \langle |\ell(x)| \rangle$ allows us to write the average distance traversed between two successive target sites as

$$\langle L \rangle = -[(\mathcal{L}_\mu - \mathcal{I})^{-1} \langle |\ell| \rangle](x_0), \tag{B.6}$$

where the mean length of a single step that starts from a point y in an interval $[\ell_0, \lambda - \ell_0]$ reads

$$\langle |\ell(y)| \rangle = \frac{(\mu - 1)\ell_0^{\mu-1}}{2} \left[\int_0^{y-\ell_0} \frac{dx}{(y-x)^{\mu-1}} + \int_{y+\ell_0}^\lambda \frac{dx}{(x-y)^{\mu-1}} \right.$$
$$\left. + y \int_{-\infty}^0 \frac{dx}{(y-x)^\mu} + (\lambda - y) \int_\lambda^\infty \frac{dx}{(x-y)^{\mu-1}} \right]. \tag{B.7}$$

The preceding equation incorporates the possibility that the flight will be truncated when it encounters one of the targets at the boundaries. In order to perform the numerical integration of Equation (B.5) in cases $Q(x) = \langle n(x) \rangle$ and $Q(x) = \langle L(x) \rangle$, we now replace the integration with a summation and the kernel $p(x - y)$ with the matrix A_{ij}, with $0 < i < M \equiv \lambda/\ell_0$, such that $A_{ii} = 0$ and

$$A_{ij} = \frac{1}{2} \left[\frac{1}{|i-j|^{\mu-1}} - \frac{1}{(|i-j|+1)^{\mu-1}} \right], \quad i \neq j. \tag{B.8}$$

The results for $\langle L \rangle$ in the nondestructive case ($x_0 = \ell_0 = 1$) are shown in Figure B.1. Notice that as $M = \lambda/\ell_0$ gets larger (i.e., for a low density of targets), the minimum in $\langle L \rangle$ (maximum in the efficiency $\eta = \langle L \rangle^{-1}$) approaches the value $\alpha = 1$ ($\mu = 2$), in agreement with the results of Chapter 10. In contrast, for $x_0 = \lambda/2$ (destructive case), it has been found [62] that the minimum in $\langle L \rangle$ tends to $\alpha = 0$ ($\mu = 1$), as expected.

It is also interesting to consider the continuous limit of the random search in which, instead of a sequence of discrete steps, the probability density of finding the Lévy searcher at point x evolves continuously with time according to the superdiffusion equation

$$\frac{\partial P(x, t)}{\partial t} = \frac{\ell_0^{\mu-1}}{t_0} \mathcal{D}_\mu P(x, t), \tag{B.9}$$

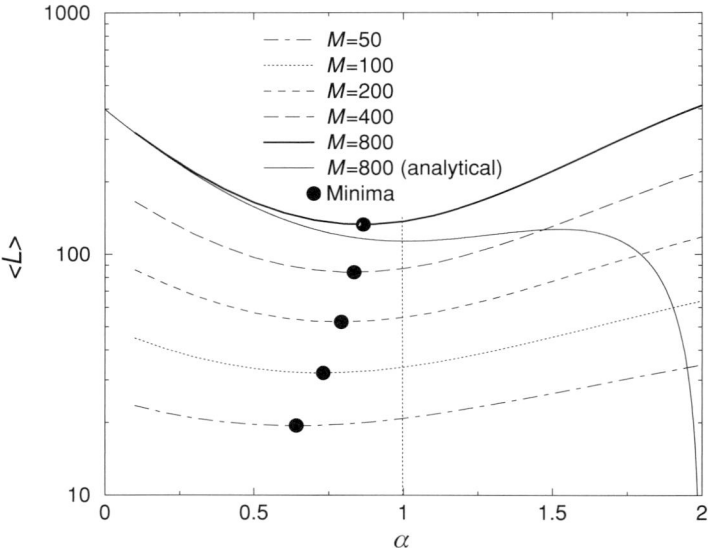

Figure B.1 Semilog plot of the numerical solutions of the average distance $\langle L \rangle$ traversed by the searcher before finding a target site at one of the boundaries in one dimension, as a function of $\alpha = \mu - 1$, for $x_0 = \ell_0 = 1$ (nondestructive search) and various values of $M \equiv \lambda/\ell_0$ (see [62]). Circles indicate the positions of the minima of $\langle L \rangle$ (maximum efficiency η), which shift toward $\alpha = 1$ ($\mu = 2$) as M increases. The analytical continuous limit approximation, Equation (B.15), is also shown for $M = 800$. Note that the optimal value $\alpha = 1$ corresponds to $\mu = 2$, explaining why Lévy walks optimize random searches (see Chapter 10).

where t_0 is the constant duration of each flight, $\ell_0^{\mu-1}/t_0$ is the fractional analogue of the diffusion coefficient, and

$$\mathcal{D}_\mu \equiv \lim_{\ell_0 \to 0} \ell_0^{-\mu+1} [\mathcal{L}_\mu(\ell_0) - \mathcal{I}]. \tag{B.10}$$

Formal substitution of Equation (B.10) into Equation (B.4) leads to the following closed-form expression for the average time spent by the continuous Lévy flight process before absorption:

$$\langle t \rangle = t_0 \langle n \rangle = \frac{t_0}{\ell_0^{\mu-1}} [\mathcal{D}_\mu^{-1} h](x_0) = \frac{t_0}{\ell_0^{\mu-1}} g(x_0), \tag{B.11}$$

where $g(x)$ satisfies

$$\mathcal{D}_\mu g(x) = h(x) = -1, \tag{B.12}$$

with boundary conditions $g(0) = g(\lambda) = 0$. Equation (B.12) belongs to a class of generalized Abel integral equations with a Riesz fractional kernel, whose solution for $1 < \mu \leq 3$

implies

$$g(x) = \frac{\sin[\pi(\mu - 1)/2]}{\pi(\mu - 1)/2}[(\lambda - x)x]^{(\mu-1)/2} . \tag{B.13}$$

For $\mu > 3$, Equation (B.9) should be replaced by the standard diffusion equation, with diffusion coefficient $D = (\mu - 1)\ell_0^2/[2t_0(\mu - 3)]$, so that $\langle t \rangle = x(\lambda - x)/(2D)$.

In the limit $\ell_0 \to 0$, the formal expansion of $(\mathcal{L}_\mu - \mathcal{I})^{-1}$ allows one to identify $\ell_0^{-\mu+1} g(x_0)$ as the first term of the expansion of the average number of flights $\langle n \rangle$ of the discrete process. Because the whole formal expansion may not converge, its first term is then defined as the average number of flights for the continuous process:

$$n_\mu(x_0) \equiv \ell_0^{-\mu+1} g(x_0) = \frac{\sin[\pi(\mu - 1)/2]}{\pi(\mu - 1)/2} \left[\frac{(\lambda - x_0)x_0}{\ell_0^2} \right]^{(\mu-1)/2} . \tag{B.14}$$

Note that this result is in agreement with Equations (10.4) and (10.5), respectively, for $x_0 = \lambda/2$ (destructive search) and $x_0 = \ell_0 = r_v \ll \lambda$ (nondestructive search), which were derived in Chapter 10 using scaling arguments. In particular, for $x_0 = \lambda/2$, it has been found [62] that there is good convergence of the scaled average number of flights of the discrete process, $\langle n \rangle (M/2)^{-\mu+1}$, with $\langle n \rangle$, as in Equation (B.4), to the analytical continuous limit expression (Equation (B.14)).

Finally, in analogy with the previous procedure, an analytical expression for the continuous limit approximation of the average total path length before absorption can be found in terms of hypergeometric functions F:

$$L_\mu(x_0) = \frac{\lambda(3 - \mu)}{2(2 - \mu)} \left[1 - 4 \frac{\psi_\mu(z) + \psi_\mu(1 - z)}{(\mu - 1)(\mu + 1)B((\mu - 1)/2, (\mu - 1)/2)} \right]$$
$$+ \frac{2\lambda M^{\mu-2} \sin(\pi(\mu - 1)/2)(z - z^2)^{(\mu-1)/2}}{\pi(\mu - 2)} , \tag{B.15}$$

where $z = x_0/\lambda$, $B(a, b) = \Gamma(a)\Gamma(b)/\Gamma(a + b)$ is the Euler B function,

$$\psi_\mu(z) = F\left(2 - \frac{(\mu - 1)}{2}, \frac{(\mu - 1)}{2}, \frac{(\mu - 1)}{2} + 2, z \right) z^{(\mu-1)/2+1} , \tag{B.16}$$

and

$$F(a, b, c, x) = \frac{\Gamma(c)}{\Gamma(a)\Gamma(b)} \sum_{n=0}^{\infty} \frac{\Gamma(n + a)\Gamma(n + b)x^n}{\Gamma(n + 1)\Gamma(n + c)} . \tag{B.17}$$

Note that in the nondestructive continuous case, L_μ presents two minima, at $\mu \to 2$ and $\mu = 3$ (Figure B.1). The former is consistent with the maximum in the efficiency η discussed in Chapter 10, whereas the latter does not exist in the discrete search process. However, the approximations made in the limit $\ell_0 \to 0$ break down in the vicinity of the absorbing boundaries, $x_0 = 0$ or $x_0 = \lambda$, when $\mu \to 3$.

References

[1] Adamic, L. A., Lukose, R. M., Puniyani, A. R., and Huberman, B. A. 2001. Search in power-law networks. *Physical Review E*, **64**, 046135.

[2] Aldana, M., Dossetti, V., Huepe, C., Kenkre, V. M., and Larralde, H. 2007. Phase transitions in systems of self-propelled agents and related network models. *Physical Review Letters*, **98**, 095702.

[3] Allegrini, P., Bellazzini, J., Bramanti, G., *et al.* 2002. Scaling breakdown: A signature of aging. *Physical Review E*, **66**, 015101.

[4] Alves-Pereira, A. R., Nunes-Pereira, E. J., Martinho, J. M. G., and Berberan-Santos, M. N. 2007. Photonic superdiffusive motion in resonance line radiation trapping: Partial frequency redistribution effects. *Journal of Chemical Physics*, **126**, 154505.

[5] Amaral, L. A. N., Cizeau, P., Gopikrishnan, P., *et al.* 1999. Econophysics: Can statistical physics contribute to the science of economics? *Computer Physics Communications*, **122**, 145–152.

[6] Amaral, L. A. N., Scala, A., Barthelemy, M., and Stanley, H. E. 2000. Classes of small-world networks. *Proceedings of the National Academy of Sciences of the United States of America*, **97**, 11 149–11 152.

[7] Angelico, R., Ceglie, A., Olsson, U., Palazzo, G., and Ambrosone, L. 2006. Anomalous surfactant diffusion in a living polymer system. *Physical Review E*, **74**, 031403.

[8] Anteneodo, C., and Morgado, W. A. M. 2007. Critical scaling in standard biased random walks. *Physical Review Letters*, **99**, 180602.

[9] Anteneodo, C., Dias, J. C., and Mendes, R. S. 2006. Long-time behavior of spreading solutions of Schrodinger and diffusion equations. *Physical Review E*, **73**, 051105.

[10] Ares, J. O., Dignani, J., and Bertiller, M. B. 2007. Cost analysis of remotely sensed foraging paths in patchy landscapes with plant anti-herbivore defenses (Patagonia, Argentina). *Landscape Ecology*, **22**, 1291–1301.

[11] Atkinson, R. P. D., Rhodes, C. J., Macdonald, D. W., and Anderson, R. M. 2002. Scale-free dynamics in the movement patterns of jackals. *Oikos*, **98**, 134–140.

[12] Austin, D., Bowen, W. D., and McMillan, J. I. 2004. Intraspecific variation in movement patterns: Modeling individual behaviour in a large marine predator. *Oikos*, **105**, 15–30.

[13] Baeumer, B., Kovács, M., and Meerschaert, M. M. 2007. Fractional reproduction-dispersal equations and heavy tail dispersal kernels. *Bulletin of Mathematical Biology*, **69**, 2281–2297.

[14] Bak, P. 1996. *How Nature Works: The Science of Self-Organized Criticality.* New York: Copernicus.

[15] Barabási, A. L. 2003. *Linked: How Everything Is Connected to Everything Else and What It Means for Business, Science, and Everyday Life.* New York: Plume.

[16] Barabási, A. L., and Stanley, H. E. 1995. *Fractal Concepts in Surface Growth.* Cambridge: Cambridge University Press.

[17] Bartumeus, F. 2007. Lévy processes in animal movement: An evolutionary hypothesis. *Fractals – Complex Geometry Patterns and Scaling in Nature and Society,* **15,** 151–162.

[18] Bartumeus, F., and Catalan, J. 2009. Optimal search behavior and classic foraging theory. *Journal of Physics A,* **42,** 434002.

[19] Bartumeus, F., Catalan, J., Fulco, U. L., Lyra, M. L., and Viswanathan, G. M. 2002. Optimizing the encounter rate in biological interactions: Lévy versus Brownian strategies. *Physical Review Letters,* **88,** 097901.

[20] Bartumeus, F., Peters, F., Pueyo, S., Marrase, C., and Catalan, J. 2003. Helical Lévy walks: Adjusting searching statistics to resource availability in microzooplankton. *Proceedings of the National Academy of Sciences of the United States of America,* **100,** 12 771–12 775.

[21] Bartumeus, F., da Luz, M. G. E., Viswanathan, G. M., and Catalan, J. 2005. Animal search strategies: A quantitative random-walk analysis. *Ecology,* **86,** 3078–3087.

[22] Bartumeus, F., Fernandez, P., da Luz, M. G. E., *et al.* 2008. Superdiffusion and encounter rates in diluted, low dimensional worlds. *European Physical Journal – Special Topics,* **157,** 157–166.

[23] Bartumeus, F., Catalan, J., Viswanathan, G. M., Raposo, E. P., and da Luz, M. G. E. 2008. The influence of turning angles on the success of non-oriented animal searches. *Journal of Theoretical Biology,* **252,** 43–55.

[24] Bartumeus, F., Giuggioli, L., Louzao, M., *et al.* 2010. Fishery discards impact on seabird movement patterns at regional scales. *Current Biology,* **20,** 215–222.

[25] Beggs, J. M., and Plenz, D. 2003. Neuronal avalanches in neocortical circuits. *Journal of Neuroscience,* **23,** 11167–11177.

[26] Belik, V. V., and Brockmann, D. 2007. Accelerating random walks by disorder. *New Journal of Physics,* **9,** 54.

[27] Benhamou, S. 2004. How to reliably estimate the tortuosity of an animal's path: Straightness, sinuosity, or fractal dimension? *Journal of Theoretical Biology,* **229,** 209–220.

[28] Benhamou, S. 2007. How many animals really do the Lévy walk? *Ecology,* **88,** 1962–1969.

[29] Benhamou, S. 2008. How many animals really do the Lévy walk? Reply. *Ecology,* **89,** 2351–2352.

[30] Bénichou, O., Coppey, M., Moreau, M., Suet, P.-H., and Voituriez, R. 2005. A stochastic model for intermittent search strategies. *Journal of Physics – Condensed Matter,* **17,** S4275–S4286.

[31] Bénichou, O., Coppey, M., Moreau, M., Suet, P.-H., and Voituriez, R. 2005. A stochastic theory for the intermittent behaviour of foraging animals. *Physica A,* **356,** 151–156.

[32] Bénichou, O., Coppey, M., Moreau, M., and Voituriez, R. 2006. Intermittent search strategies: When losing time becomes efficient. *Europhysics Letters,* **75,** 349–354.

[33] Bénichou, O., Loverdo, C., Moreau, M., and Voituriez, R. 2006. Two-dimensional intermittent search processes: An alternative to Lévy flight strategies. *Physical Review E,* **74,** 020102.

[34]   Bénichou, O., Loverdo, C., Moreau, M., and Voituriez, R. 2007. A minimal model of intermittent search in dimension two. *Journal of Physics – Condensed Matter*, **19**, 065141.

[35]   Berg, H. C. 1993. *Random Walks in Biology*. Princeton, NJ: Princeton University Press.

[36]   Berkolaiko, G., and Havlin, S. 1997. Territory covered by *N* Lévy flights on *d*-dimensional lattices. *Physical Review E*, **55**, 1395–1400.

[37]   Berkolaiko, G., and Havlin, S. 1998. Number of distinct sites visited by Lévy flights injected into a *d*-dimensional lattice. *Physical Review E*, **57**, 2549–2552.

[38]   Bertiller, M. B., and Ares, J. O. 2008. Sheep spatial grazing strategies at the arid Patagonian Monte, Argentina. *Rangeland Ecology and Management*, **61**, 38–47.

[39]   Bertrand, S., Burgos, J. M., Gerlotto, F., and Atiquipa, J. 2005. Lévy trajectories of Peruvian purse-seiners as an indicator of the spatial distribution of anchovy (*Engraulis ringens*). *ICES Journal of Marine Science*, **62**, 477–482.

[40]   Bertrand, S., Bertrand, A., Guevara-Carrasco, R., and Gerlotto, F. 2007. Scale-invariant movements of fishermen: The same foraging strategy as natural predators. *Ecological Applications*, **17**, 331–337.

[41]   Bovet, J., and Bovet, P. 1993. Computer-simulations of rodent homing behaviour, using a probabilistic model. *Journal of Theoretical Biology*, **161**, 145–156.

[42]   Bovet, P., and Benhamou, S. 1988. Spatial analysis of animals' movements using a correlated random walk model. *Journal of Theoretical Biology*, **131**, 419–433.

[43]   Bovet, P., and Benhamou, S. 1991. Optimal sinuosity in central place foraging movements. *Animal Behaviour*, **42**, 57–62.

[44]   Boyer, D., and Larralde, H. 2005. Looking for the right thing at the right place: Phase transition in an agent model with heterogeneous spatial resources. *Complexity*, **10**, 52–55.

[45]   Boyer, D., and Lopez-Corona, O. 2009. Self-organization, scaling and collapse in a coupled automaton model of foragers and vegetation resources with seed dispersal. *Journal of Physics A*, **42**, 434014.

[46]   Boyer, D., Miramontes, O., Ramos-Fernández, G., Mateos, J. L., and Cocho, G. 2004. Modeling the searching behavior of social monkeys. *Physica A*, **342**, 329–335.

[47]   Boyer, D., Ramos-Fernández, G., Miramontes, O., *et al.* 2006. Scale-free foraging by primates emerges from their interaction with a complex environment. *Proceedings of the Royal Society B*, **273**, 1743–1750.

[48]   Boyer, D., Miramontes, O., and Ramos-Fernández, G. 2007. Evidence for biological Lévy flights stands. arXiv:0802.1762v1 [q-bio.PE].

[49]   Boyer, D., Miramontes, O., and Larralde, H. 2009. Levy-like behaviour in deterministic models of intelligent agents exploring heterogeneous environments. *Journal of Physics A*, **42**, 434015.

[50]   Bradshaw, C. J. A., Sims, D. W., and Hays, G. C. 2007. Measurement error causes scale-dependent threshold erosion of biological signals in animal movement data. *Ecological Applications*, **17**, 628–638.

[51]   Brantingham, P. J. 2006. Measuring forager mobility. *Current Anthropology*, **47**, 435–459.

[52]   Brockmann, D. 2008. Anomalous diffusion and the structure of human transportation networks. *European Physical Journal – Special Topics*, **157**, 173–189.

[53]   Brockmann, D., and Geisel, T. 2000. The ecology of gaze shifts. *Neurocomputing*, **32**, 643–650.

[54] Brockmann, D., and Hufnagel, L. 2007. Front propagation in reaction-superdiffusion dynamics: Taming Lévy flights with fluctuations. *Physical Review Letters*, **98**, 178301.

[55] Brockmann, D., and Sokolov, I. M. 2002. Lévy flights in external force fields: From models to equations. *Chemical Physics*, **284**, 409–421.

[56] Brockmann, D., Hufnagel, L., and Geisel, T. 2006. The scaling laws of human travel. *Nature*, **439**, 462–465.

[57] Brown, C. T., Witschey, W. R. T., and Liebovitch, L. S. 2005. The broken past: Fractals in archaeology. *Journal of Archaeological Method and Theory*, **12**, 37–78.

[58] Brown, C. T., Liebovitch, L. S., and Glendon, R. 2007. Lévy flights in Dobe Ju/'hoansi foraging patterns. *Human Ecology*, **35**, 129–138.

[59] Buchanan, M. 2008. Ecological modelling: The mathematical mirror to animal nature. *Nature*, **453**, 714–716.

[60] Buendía, G. M., Viswanathan, G. M., and Kenkre, V. M. 2008. Multifractality of random walks in the theory of vehicular traffic. *Physical Review E*, **78**, 056110.

[61] Buiatti, M., Papo, D., Baudonniere, P.-M., and Van Vreeswijk, C. 2007. Feedback modulates the temporal scale-free dynamics of brain electrical activity in a hypothesis testing task. *Neuroscience*, **146**, 1400–1412.

[62] Buldyrev, S. V., Havlin, S., Kazakov, A. Y., *et al.* 2001. Average time spent by Lévy flights and walks on an interval with absorbing boundaries. *Physical Review E*, **6404**, 041108.

[63] Buldyrev, S. V., Gitterman, M., Havlin, S., *et al.* 2001. Properties of Lévy flights on an interval with absorbing boundaries. *Physica A*, **302**, 148–161.

[64] Bunde, A., and Havlin, S. (eds.). 1991. *Fractals and Disordered Systems*. Berlin: Springer.

[65] Bunde, A., and Havlin, S. (eds.). 1994. *Fractals in Science*. Berlin: Springer.

[66] Bunimovich, L. A. 2003. Walks in rigid environments: Symmetry and dynamics. *Asterisque*, **286**, 231–248.

[67] Bunimovich, L. A. 2004. Deterministic walks in random environments. *Physica D*, **187**, 20–29.

[68] Burrell, K. H., Isely, J. J., Bunnell, D. B., Van Lear, D. H., and Dolloff, C. A. 2000. Seasonal movement of brown trout in a southern Appalachian river. *Transactions of the American Fisheries Society*, **129**, 1373–1379.

[69] Cabrera, J. L., and Milton, J. G. 2004. Human stick balancing: Tuning Lévy flights to improve balance control. *Chaos*, **14**, 691–698.

[70] Cabrera, J. L., Bormann, R., Eurich, C., Ohira, T., and Milton, J. 2004. State-dependent noise and human balance control. *Fluctuation and Noise Letters*, **4**, L107–L117.

[71] Cabrera, J. L., Luciani, C., and Milton, J. 2006. Neural control on multiple time scales: Insights from human stick balancing. *Condensed Matter Physics*, **9**, 373–383.

[72] Cao, X.-Q., Zend, J., and Yan, H. 2009. Physical signals for protein-DNA recognition. *Physical Biology*, **6**, 036012.

[73] Cascetta, E., and Russo, F. 1997. Calibrating aggregate travel demand models with traffic counts: Estimators and statistical performance. *Transportation*, **24**, 271–293.

[74] Chambers, R., Bickel, W. K., and Potenza, M. N. 2007. A scale-free systems theory of motivation and addiction. *Neuroscience and Biobehavioral Reviews*, **31**, 1017–1045.

[75] Charnov, E. L. 1976. Optimal foraging: Attack strategy of a mantid. *American Naturalist*, **110**, 141–151.

[76] Charnov, E. L. 1976. Optimal foraging: The marginal value theorem. *Theoretical Population Biology*, **9**, 129–136.

[77] Chechkin, A. V., Metzler, R., Gonchar, V. Y., Klafter, J., and Tanatarov, L. V. 2003. First passage and arrival time densities for Lévy flights and the failure of the method of images. *Journal of Physics A*, **36**, L537–L544.

[78] Chechkin, A. V., Gonchar, V. Y., Klafter, J., Metzler, R., and Tanatarov, L. V. 2004. Lévy flights in a steep potential well. *Journal of Statistical Physics*, **115**, 1505–1535.

[79] Chechkin, A. V., Gonchar, V. Y., Klafter, J., and Metzler, R. 2005. Natural cutoff in Lévy flights caused by dissipative nonlinearity. *Physical Review E*, **72**, 010101.

[80] Chechkin, A. V., Gonchar, V. Y., Gorenflo, R., Korabel, N., and Sokolov, I. M. 2008. Generalized fractional diffusion equations for accelerating subdiffusion and truncated Lévy flights. *Physical Review E*, **78**, 021111.

[81] Chechkin, A. V., Metzler, R., Klafter, J., and Gonchar, V. Y. 2008. Introduction to the theory of Lévy flights. In: *Anomalous Transport*. Berlin: Wiley-VCH, 129–162.

[82] Cheung, A., Zhang, S., Stricker, C., and Srinivasan, M. V. 2007. Animal navigation: The difficulty of moving in a straight line. *Biological Cybernetics*, **97**, 47–61.

[83] Cheung, A., Zhang, S., Stricker, C., and Srinivasan, M. V. 2008. Animal navigation: General properties of directed walks. *Biological Cybernetics*, **99**, 197–217.

[84] Codling, E. A., Plank, M. J., and Benhamou, S. 2008. Random walk models in biology. *Journal of the Royal Society Interface*, **5**, 813–834.

[85] Cole, B. J. 1995. Fractal time in animal behaviour: The movement activity of *Drosophila*. *Animal Behavior*, **50**, 1317–1324.

[86] Condat, C. A., Rangel, J., and Lamberti, P. W. 2002. Anomalous diffusion in the nonasymptotic regime. *Physical Review E*, **65**, 026138.

[87] Coppey, M., Bénichou, O., Klafter, J., Moreau, M., and Oshanin, G. 2004. Catalytic reactions with bulk-mediated excursions: Mixing fails to restore chemical equilibrium. *Physical Review E*, **69**, 036115.

[88] Cressoni, J. C., da Silva, M. A. A., and Viswanathan, G. M. 2007. Amnestically induced persistence in random walks. *Physical Review Letters*, **98**, 070603.

[89] Crook, J. H. (ed.). 1970. *Social Behaviour in Birds and Mammals*. London: Academic Press.

[90] da Luz, M. G. E., Buldyrev, S. V., Havlin, S., *et al.* 2001. Improvements in the statistical approach to random Lévy flight searches. *Physica A*, **295**, 89–92.

[91] da Luz, M. G. E., Grosberg, A., Raposo, E. P., and Viswanathan, G. M. (eds.). 2009. The random search problem: Trends and perspectives [special issue] *Journal of Physics A*, **42**, no. 43.

[92] Dahirel, V., Paillusson, F., Jardat, M., Barbi, M., and Victor, J.-M. 2009. Nonspecific DNA-protein interaction: Why proteins can diffuse along DNA. *Physical Review Letters*, **102**, 228101.

[93] Dahl, O. C. 1991. *Migration from Kalimantan to Madagascar*. Oslo: Norwegian University Press.

[94] Dai, X., Shannon, G., Slotow, R., Page, B., and Duffy, K. J. 2007. Short-duration daytime movements of a cow herd of African elephants. *Journal of Mammalogy*, **88**, 151–157.

[95] Davies, P. 1989. *The New Physics*. Cambridge: Cambridge University Press.

[96] Dawkins, R. 1986. *The Blind Watchmaker*. Harlow, UK: Longman.

[97] De Knegt, H. J., Hengeveld, G. M., van Langevelde, F., de Boer, W. F., and Kirkman, K. P. 2007. Patch density determines movement patterns and foraging efficiency of large herbivores. *Behavioral Ecology*, **18**, 1065–1072.

[98] Denisov, S. I., Horsthemke, W., and Haenggi, P. 2008. Steady-state Lévy flights in a confined domain. *Physical Review E*, **77**, 061112.

[99] Diamond, J. M. 2000. Linguistics: Taiwan's gift to the world. *Nature*, **403**, 709–710.

[100] Doi, M. 1989. *Introduction to Polymer Physics*. Oxford: Oxford University Press.

[101] Doran, E. B. 1981. *Wangka: Austronesian Canoe Origins*. College Station: Texas A&M University Press.

[102] Dubkov, A., and Spagnolo, B. 2007. Langevin approach to Lévy flights in fixed potentials: Exact results for stationary probability distributions. *Acta Physica Polonica B*, **38**, 1745–1758.

[103] Dybiec, B. 2008. Random strategies of contact tracking. *Physica A*, **387**, 4863–4870.

[104] Ebert, L. A., Schaerli, P., and Moser, B. 2005. Chemokine-mediated control of T cell traffic in lymphoid and peripheral tissues. *Molecular Immunology*, **42**, 799–809.

[105] Edwards, A. M., Phillips, R. A., Watkins, N. W., *et al.* 2007. Revisiting Lévy flight search patterns of wandering albatrosses, bumblebees and deer. *Nature*, **449**, 1044–1048.

[106] Eilazar, I., and Klafter, J. 2003. On the extreme flights of one-sided Lévy processes. *Physica A*, **330**, 8–17.

[107] Eliazar, I., Koren, T., and Klafter, J. 2007. Searching circular DNA strands. *Journal of Physics – Condensed Matter*, **19**, 065140.

[108] Eliazar, I., Koren, T., and Klafter, J. 2008. Parallel search of long circular strands: Modeling, analysis, and optimization. *Journal of Physical Chemistry B*, **112**, 5905–5909.

[109] Emlen, J. M. 1966. The role of time and energy in food preference. *American Naturalist*, **100**, 611–617.

[110] Engbert, R. 2006. Microsaccades: A microcosm for research on oculomotor control, attention, and visual perception. *Progress in Brain Research*, **154**, 177–192.

[111] Fauchald, P. 1999. Foraging in a hierarchical patch system. *American Naturalist*, **153**, 603–613.

[112] Fauchald, P., and Tveraa, T. 2003. Using first-passage time in the analysis of area-restricted search and habitat selection. *Ecology*, **84**, 282–288.

[113] Fauchald, P., Erikstad, K. E., and Skarsfjord, H. 2000. Scale-dependent predator-prey interactions: The hierarchical spatial distribution of seabirds and prey. *Ecology*, **81**, 773–783.

[114] Faure, P., Neumeister, H., Faber, D. S., and Korn, H. 2003. Symbolic analysis of swimming trajectories reveals scale invariance and provides a model for fish locomotion. *Fractals – Complex Geometry Patterns and Scaling in Nature and Society*, **11**, 233–243.

[115] Faustino, C. L., da Silva, L. R., da Luz, M. G. E., Raposo, E. P., and Viswanathan, G. M. 2007. Search dynamics at the edge of extinction: Anomalous diffusion as a critical survival state. *Europhysics Letters*, **77**, 30002.

[116] Felisberto, M. L., da Luz, M. G. E., Bartumeus, F., Raposo, E. P., and Viswanathan, G. M. 2009. Correlated Lévy walk. In *Abstracts of the Latin American Workshop on Nonlinear Phenomena, Búzios, 05–09 October*. Curitiba-PR, Brazil: Editora Universidade Federal do Paraná, p. 27.

[117] Fenchel, T. 2004. Orientation in two dimensions: Chemosensory motile behaviour of *Euplotes vannus*. *European Journal of Protistology*, **40**, 49–55.

[118] Figueiredo, A., Gléria, I., Matsushita, R., and da Silva, S. 2006. Nonidentically distributed variables and nonlinear autocorrelation. *Physica A*, **363**, 171–180.

[119] Filfillan, S. L. 2001. An ecological study of a population of *Pseudantechinus macdonnellensis* (Marsupialia: Dasyuridae) in central Australia. II. Population dynamics and movements. *Wildlife Research*, **28**, 481–492.

[120] Fisher, M. E. 1998. Renormalization group theory: Its basis and formulation in statistical physics. *Reviews of Modern Physics*, **70**, 653.

[121] Focardi, S., Montanaro, P., and Pecchioli, E. 2009. Adaptative Lévy walks in foraging fallow deer. *PLoS ONE*, **4**, e6587.

[122] Forester, J. D., Ives, A. R., Turner, M. G., *et al.* 2007. State-space models link elk movement patterns to landscape characteristics in Yellowstone National Park. *Ecological Monographs*, **77**, 285–299.

[123] Freitas, J. F. L., and Lyra, M. L. 2003. Optimal transition rate and stochastic resonance in a bistable system driven by power-law noise. *International Journal of Modern Physics C*, **14**, 303–310.

[124] Freund, H., and Grassberger, P. 1992. The Red Queen's walk. *Physica A*, **190**, 218–237.

[125] Fritz, H., Said, S., and Weimerskirch, H. 2003. Scale-dependent hierarchical adjustments of movement patterns in a long-range foraging seabird. *Proceedings of the Royal Society B*, **270**, 1143–1148.

[126] Gale, D., Propp, J., Sutherland, S., and Troubetzkoy, S. 1995. Further travels with my ant. *Mathematical Intelligencer*, **17**, 48–56.

[127] Gao, J. B., Hu, J., Tung, W. W., and Cao, Y. H. 2006. Distinguishing chaos from noise by scale-dependent Lyapunov exponent. *Physical Review E*, **74**, 066204.

[128] Garcia, R., Moss, F., Nihongi, A., *et al.* 2007. Optimal foraging by zooplankton within patches: The case of *Daphnia*. *Mathematical Biosciences*, **207**, 165–188.

[129] Garoni, T. M., and Frankel, N. E. 2002. *d*-dimensional Lévy flights: Exact and asymptotic. *Journal of Mathematical Physics*, **43**, 5090–5107.

[130] Garoni, T. M., and Frankel, N. E. 2002. Lévy flights: Exact results and asymptotics beyond all orders. *Journal of Mathematical Physics*, **43**, 2670–2689.

[131] Gautestad, A. O., and Mysterud, I. 2005. Intrinsic scaling complexity in animal dispersion and abundance. *American Naturalist*, **165**, 44–55.

[132] Gautestad, A. O., and Mysterud, I. 2006. Complex animal distribution and abundance from memory-dependent kinetics. *Ecological Complexity*, **3**, 44–55.

[133] Geisel, T., Nierwetberg, J., and Zacherl, A. 1985. Accelerated diffusion in Josephson junctions and related chaotic systems. *Physical Review Letters*, **54**, 616.

[134] Gibbons, A. 2001. The peopling of the Pacific. *Science*, **291**, 1735–37.

[135] Giuggioli, L., Viswanathan, G. M., Kenkre, V. M., Parmenter, R. R., and Yates, T. L. 2007. Effects of finite probing windows on the interpretation of the multifractal properties of random walks. *Europhysics Letters*, **77**, 40004.

[136] Giuggioli, L., Sevilla, F. J., and Kenkre, V. M. 2009. A generalized master equation approach to modelling anomalous transport in animal movement. *Journal of Physics A*, **42**, 434004.

[137] González, M. C., Hidalgo, C. A., and Barabási, A.-L. 2008. Understanding individual human mobility patterns. *Nature*, **453**, 779–782.

[138] Goss-Custard, J. D. 1970. Feeding dispersion in some overwintering wading birds. In *Social Behavior in Birds and Mammals*, ed. J. H. Crook, pp. 3–35. London: Academic Press.

[139] Goss-Custard, J. D. 1980. Competition for food and interference among waders. *Ardea*, **68**, 31–52.

[140] Gray, R. D., Drummond, A. J., and Greenhill, S. J. 2009. Language phylogenies reveal expansion pulses and pauses in Pacific settlement. *Science*, **323**, 479–483.

[141] Grima, R. 2008. Multiscale modeling of biological pattern formation. In *Multiscale Modeling of Developmental Systems*, ed. S. Schnell *et al.* London: Academic Press, pp. 435–460. Current Topics in Developmental Biology 81.

[142] Gu, W., Regens, J. L., Beier, J. C., and Novak, R. J. 2006. Source reduction of mosquito larval habitats has unexpected consequences on malaria transmission. *Proceedings of the National Academy of Sciences of the United States of America*, **103**, 17 560–17 563.

[143] Gupta, H. M., and Campanha, J. R. 2000. The gradually truncated Lévy flight: Stochastic process for complex systems. *Physica A*, **275**, 531–543.

[144] Gupta, H. M., and Campanha, J. R. 2002. Tsallis statistics and gradually truncated Lévy flight – distribution of an economical index. *Physica A*, **309**, 381–387.

[145] Guy, A. G., Bohan, D. A., Powers, S. J., and Reynolds, A. M. 2008. Avoidance of conspecific odour by carabid beetles: A mechanism for the emergence of scale-free searching patterns. *Animal Behaviour*, **76**, 585–591.

[146] Haefner, J. W., and Crist, T. O. 1994. Spatial model of movement and foraging in harvester ants (*Pogonomyrex*) (I): The role of memory and communication. *Journal of Theoretical Biology*, **166**, 299–313.

[147] Hapca, S., Crawford, J. W., MacMillan, K., Wilson, M. J., and Young, I. M. 2007. Modelling nematode movement using time-fractional dynamics. *Journal of Theoretical Biology*, **248**, 212–224.

[148] Harnos, A., Horvath, G., Lawrence, A. B., and Vattay, G. 2000. Scaling and intermittency in animal behaviour. *Physica A*, **286**, 312–320.

[149] Havlin, S., and Benavraham, D. 1987. Diffusion in disordered media. *Advances in Physics*, **36**, 695–798.

[150] Hays, G. C., Hobson, V. J., Metcalfe, J. D., Righton, D., and Sims, D. W. 2006. Flexible foraging movements of leatherback turtles across the North Atlantic Ocean. *Ecology*, **87**, 2647–2656.

[151] Heisenberg, M. 2009. Is free will an illusion? *Nature*, **459**, 164–165.

[152] Herrmann, H. J. 1999. Statistical models for granular materials. *Physica A*, **263**, 51–62.

[153] Herrmann, H. J., and Roux, S. (eds.). 1990. *Statistical Models for the Fracture of Disordered Media*. Amsterdam: North-Holland.

[154] Hobbs, N. T. 2003. Challenges and opportunities in integrating ecological knowledge across scales. *Forest Ecology and Management*, **181**, 223–238.

[155] Hobbs, N. T., Gross, J. E., Shipley, L. A., Spalinger, D. E., and Wunder, B. A. 2003. Herbivore functional response in heterogeneous environments: A contest among models. *Ecology*, **84**, 666–681.

[156] Horn, D. J., Stairs, G. R., and Mitchell, R. D. (eds.). 1979. *Analysis of Ecological Systems*. Columbus: Ohio State University Press.

[157] Houston, A. I., and McNamara, J. M. 1999. *Models of Adaptive Behaviour: An Approach Based on State*. Cambridge: Cambridge University Press.

[158] Hoyle, M., and Cresswell, J. E. 2007. A search theory model of patch-to-patch forager movement with application to pollinator-mediated gene flow. *Journal of Theoretical Biology*, **248**, 154–163.

[159] Hu, L., Grosberg, A. Y., and Bruinsma, R. 2009. First passage time distribution for the 1D diffusion of particles with internal degrees of freedom. *Journal of Physics A*, **42**, 434011.

[160] Huet, S., Karatekin, E., Tran, V. S., *et al.* 2006. Analysis of transient behavior in complex trajectories: Application to secretory vesicle dynamics. *Biophysical Journal*, **91**, 3542–3559.

[161] Huettmann, F. 2004. Computing foraging paths for shore-birds using fractal dimensions and pecking success from footprint surveys on mudflats: An application for red-necked stints in the Moroshechnaya River Estuary, Kamchatka – Russian Far East. In *Computational Science and ITS Applications – ICCSA 2004, PT 2*, ed. M. L. Gavrilova *et al.*, pp. 1117–1120. Lecture Notes in Computer Science 3044. Berlin: Springer.

[162] Huey, R. B., and Planka, E. R. 1981. Ecological consequences of foraging mode. *Ecology*, **62**, 991–999.

[163] Humphries, N. E., Queiroz, N., Dyer, J. R. M., *et al.* 2010. Environmental context explains Lévy and Brownian movement patterns of marine predators. *Nature*, **465**, 1066–1069.

[164] Hurst, H. E., Black, R. P., and Simaika, Y. M. 1965. *Long-Term Storage: An Experimental Study*. London: Constable.

[165] Imkeller, P., and Pavlyukevich, I. 2006. Lévy flights: Transitions and meta-stability. *Journal of Physics A*, **39**, L237–L246.

[166] Jennings, H. D., Ivanov, P. C., Martins, A. M., Silva, P. C., and Viswanathan, G. M. 2004. Variance fluctuations in nonstationary time series: A comparative study of music genres. *Physica A*, **336**, 585.

[167] Jespersen, S., Metzler, R., and Fogedby, H. C. 1999. Lévy flights in external force fields: Langevin and fractional Fokker-Planck equations and their solutions. *Physical Review E*, **59**, 2736–2745.

[168] Johnson, C. J., Boyce, M. S., Mulders, R., *et al.* 2004. Quantifying patch distribution at multiple spatial scales: Applications to wildlife-habitat models. *Landscape Ecology*, **19**, 869–882.

[169] Johnson, S. N., Crawford, J. W., Gregory, P. J., *et al.* 2007. Non-invasive techniques for investigating and modelling root-feeding insects in managed and natural systems. *Agricultural and Forest Entomology*, **9**, 39–46.

[170] Kabatiansky, G., and Oshanin, G. 2009. Finding passwords by random walks: how long does it take? *Journal of Physics A*, **42**, 434016.

[171] Kamil, A. C., and Sargent, T. D. (eds.). 1981. *Foraging Behavior: Ecological, Ethological, and Psychological Approaches*. New York: Garland.

[172] Kamil, A. C., Drebs, J. R., and Pulliam, H. R. (eds.). 1987. *Foraging Behavior*. New York: Plenum Press.

[173] Kareiva, P. M., and Shigesada, N. 1983. Analyzing insect movement as a correlated random walk. *Oecologia*, **56**, 234–238.

[174] Katori, H., Schlipf, S., and Walther, H. 1997. Anomalous dynamics of a single ion in an optical lattice. *Physical Review Letters*, **79**, 2221–2224.

[175] Kenkre, V. M. 1977. The generalized master equation and its applications. In *Statistical Mechanics and Statistical Methods in Theory and Application*, ed. U. Landman, pp. 441–461. New York: Plenum.

[176] Kenkre, V. M. 2003. Memory formalism, nonlinear techniques, and kinetic equation approaches. In *Modern Challenges in Statistical Mechanics: Patterns, Noice, and the Interplay of Nonlinearity and Complexity*, ed. V. M. Kenkre and K. Lindenberg, pp. 63–102. Melville, NY: American Institute of Physics.

[177] Kenkre, V. M. 2007. Analytic formulation, exact solutions, and generalizations of the elephant and the Alzheimer random walks. arXiv:0708.0034v2 [cond-mat.stat-mech].

[178] Kenkre, V. M., and Knox, R. S. 1974. Generalized-master-equation theory of excitation transfer. *Physical Review B*, **9**, 5279–5290.

[179] Kenkre, V. M., and Lindenberg, K. (eds.). 2003. *Modern Challenges in Statistical Mechanics: Patterns, Noise and the Interplay of Nonlinearity and Complexity*

[Proceedings of the Pan American Advanced Studies Institute, Bariloche, Argentina]. Melville, NY: American Institute of Physics.

[180] Kenkre, V. M., Montroll, E. W., and Shlesinger, M. F. 1973. Generalized master equations for continuous-time random walks. *Journal of Statistical Physics*, **9**, 45–50.

[181] Khinchin, A. I. 1949. *Mathematical Foundations of Statistical Mechanics*. New York: Courier Dover.

[182] Kinouchi, O., Martinez, A. S., Lima, G. F., Lourenço, G. M., and Risau-Gusman, S. 2002. Deterministic walks in random networks: An application to thesaurus graphs. *Physica A*, **315**, 665–676.

[183] Kiorboe, T., Grossart, H. P., Ploug, H., and Tang, K. 2002. Mechanisms and rates of bacterial colonization of sinking aggregates. *Applied and Environmental Microbiology*, **68**, 3996–4006.

[184] Kish, D. 2009. Echo vision: The man who sees with sound. *New Scientist*, **202** (2703), 31–33.

[185] Klafter, J., Blumen, A., and Shlesinger, M. F. 1987. Stochastic pathway to anomalous diffusion. *Physical Review A*, **35**, 3081.

[186] Kleiber, M. 1947. Body size and metabolic rate. *Physiological Reviews*, **27**, 511–541.

[187] Kleinberg, J. M. 2000. Navigation in a small world. *Nature*, **406**, 845.

[188] Kölzsch, A., and Blasius, B. 2008. Theoretical approaches to bird migration. *European Physical Journal – Special Topics*, **157**, 191–208.

[189] Korobkova, E., Emonet, T., Vilar, J. M. G., Shimizu, T. S., and Cluzel, P. 2004. From molecular noise to behavioural variability in a single bacterium. *Nature*, **428**, 574–578.

[190] Krebs, J. R. 1978. Optimal foraging: decision rules for predators. In *Behavioural Ecology: An evolutionary Approach*, ed. J. R. Krebs and N. B. Davies, pp. 34–63. Oxford: Blackwell.

[191] Krebs, J. R., and Davies, N. B. (eds.). 1978. *Behavioral Ecology: An Evolutionary Approach*. Oxford: Blackwell.

[192] Krebs, J. R., Kacelnik, A., and Taylor, P. 1978. Test of optimal sampling by foraging great tits. *Nature*, **275**, 27–31.

[193] Kumar, N., Viswanathan, G. M., and Kenkre, V. M. 2009. Hurst exponents for interacting random walkers obeying nonlinear Fokker-Planck equations. *Physica A*, **388**, 3687–3694.

[194] Kutner, R. 1999. Hierarchical spatio-temporal coupling in fractional wanderings. (I) – Continuous-time Weierstrass flights. *Physica A*, **264**, 84–106.

[195] Kutner, R., and Maass, P. 1998. Lévy flights with quenched noise amplitudes. *Journal of Physics A*, **31**, 2603–2609.

[196] Lacasta, A. M., Sancho, J. M., Romero, A. H., Sokolov, I. M., and Lindenberg, K. 2004. From subdiffusion to superdiffusion of particles on solid surfaces. *Physical Review E*, **70**, 051104.

[197] Lamine, K., Lambin, M., and Alauzet, C. 2005. Effect of starvation on the searching path of the predatory bug *Deraecoris lutescens*. *BioControl*, **50**, 717–727.

[198] Landman, U. (ed.). 1977. *Statistical Mechanics and Statistical Methods in Theory and Application*. New York: Plenum.

[199] Larralde, H., Trunfio, P., Havlin, S., Stanley, H. E., and Weiss, G. H. 1992. Territory covered by *N* diffusing particles. *Nature*, **355**, 423–426.

[200] Latora, V., Rapisarda, A., and Tsallis, C. 2001. Non-Gaussian equilibrium in a long-range Hamiltonian system. *Physical Review E*, **64**, 056134.

[201] Lauzon-Guay, J. S., Scheibling, R. E., and Barbeau, M. A. 2006. Movement patterns in the green sea urchin, *Strongylocentrotus droebachaensis*. *Journal of the Marine Biological Association of the United Kingdom*, **86**, 167–174.

[202] Lee, S. H., Bardunias, P., and Su, N. Y. 2007. Optimal length distribution of termite tunnel branches for efficient food search and resource transportation. *Biosystems*, **90**, 802–807.

[203] Lee, S. H., Bardunias, P., and Su, N. Y. 2008. Two strategies for optimizing the food encounter rate of termite tunnels simulated by a lattice model. *Ecological Modelling*, **213**, 381–388.

[204] Leggett, K. E. A. 2006. Home range and seasonal movement of elephants in the Kunene region, northwestern Namibia. *African Zoology*, **41**(1), 17–36.

[205] Levandowsky, M., Klafter, J., and White, B. S. 1988. Feeding and swimming behavior in grazing zooplankton. *Journal of Protozoology*, **35**, 243–246.

[206] Levandowsky, M., Klafter, J., and White, B. S. 1988. Swimming behavior and chemosensory responses in the protistan microzooplankton as a function of the hydrodynamic regime. *Bulletin of Marine Science*, **43**, 758–763.

[207] Levandowsky, M., White, B. S., and Schuster, F. L. 1997. Random movements of soil amebas. *Acta Protozoologica*, **36**, 237–248.

[208] Levina, A., Herrmann, J. M., and Geisel, T. 2007. Dynamical synapses causing self-organised criticality in neural networks. *Nature Physics*, **3**, 857–860.

[209] Lévy, P. 1937. *Théorie de l'addition des Variables Aléatoires*. Paris: Gauthier-Villars.

[210] Libet, B. 2004. *Mind Time: The Temporal Factor in Consciousness*. Cambridge: Harvard University Press.

[211] Lima, G. F., Martinez, A. S., and Kinouchi, O. 2001. Deterministic walks in random media. *Physical Review Letters*, **8701**, 010603.

[212] Lomholt, M. A., Ambjornsson, T., and Metzler, R. 2005. Optimal target search on a fast-folding polymer chain with volume exchange. *Physical Review Letters*, **95**, 260603.

[213] Lomholt, M. A., Koren, T., Metzler, R., and Klafter, J. 2008. Lévy strategies in intermittent search processes are advantageous. *Proceedings of the National Academy of Sciences of the United States of America*, **105**, 11 055–11 059.

[214] Lomholt, M. A., van den Broek, B., Kalisch, S.-M. J., Wuite, G. J. L., and Metzler, R. 2009. Facilitated diffusion with DNA coiling. *Proceedings of the National Academy of Sciences of the United States of America*, **106**, 8204–8208.

[215] Luque, B., Miramontes, O., and Lacasa, L. 2008. Number theoretic example of scale-free topology inducing self-organized criticality. *Physical Review Letters*, **101**, 158702.

[216] Lutz, C., Kollmann, M., and Bechinger, C. 2004. Single-file diffusion of colloids in one-dimensional channels. *Physical Review Letters*, **93**, 026001.

[217] Maass, P., and Scheffler, F. 2002. Lévy field distributions and anomalous spin relaxation in disordered magnetic systems. *Physica A*, **314**, 200–207.

[218] MacArthur, R. H. 1972. *Geographical Ecology: Patterns in the Distribution of Species*. New York: Harper and Row.

[219] MacArthur, R. H., and Pianka, E. R. 1966. On the optimal use of a patchy environment. *American Naturalist*, **100**, 603–609.

[220] Mandelbrot, B. B. 1982. *The Fractal Geometry of Nature*. San Francisco: Freeman.

[221] Mantegna, R. N., and Stanley, H. E. 1994. Stochastic process with ultraslow convergence to a Gaussian: The truncated Lévy flight. *Physical Review Letters*, **73**, 2946.

[222] Mantegna, R. N., and Stanley, H. E. 2000. *An Introduction to Econophysics*. Cambridge: Cambridge University Press.

[223] Marchal, P., Poos, J.-J., and Quirijns, F. 2007. Linkage between fishers' foraging, market and fish stocks density: Examples from some North Sea fisheries. *Fisheries Research*, **83**, 33–43.

[224] Mårell, A., Ball, J. P., and Hofgaard, A. 2002. Foraging and movement paths of female reindeer: Insights from fractal analysis, correlated random walks, and Lévy flights. *Canadian Journal of Zoology*, **80**, 854–865.

[225] Martin, J. R., Faure, P., and Ernst, R. 2001. The power law distribution for walking-time intervals correlates with the ellipsoid-body in *Drosophila*. *Journal of Neurogenetics*, **15**, 205–219.

[226] Martinez, A. S., Kinouchi, O., and Risau-Gusman, S. 2004. Exploratory behavior, trap models, and glass transitions. *Physical Review E*, **69**, 017101.

[227] Martinez-Conde, S., Macknik, S. L., Troncoso, X. G., and Dyar, T. A. 2006. Microsaccades counteract visual fading during fixation. *Neuron*, **49**, 297–305.

[228] Martinez-Conde, S., Macknik, S. L., Martinez, L. M., Alonso, J. M., and Tse, P. U. (eds.). 2006. *Visual Perception, Part 1, Fundamentals of Vision: Low and Mid-level Processes in Perception*. Progress in Brain Research 154. Amsterdam: Elsevier.

[229] Martins, M. L., Ceotto, G., Alves, S. G., *et al.* 2000. Cellular automata model for citrus variegated chlorosis. *Physical Review E*, **62**, 7024–7030.

[230] Mashanov, G. I., and Molloy, J. E. 2007. Automatic detection of single fluorophores in live cells. *Biophysical Journal*, **92**, 2199–2211.

[231] Mason, O., and Verwoerd, M. 2007. Graph theory and networks in biology. *IET Systems Biology*, **1**, 89–119.

[232] Masson, J.-B., Bechet, M. B., and Vergassola, M. 2009. Chasing information to search in random environments. *Journal of Physics A*, **42**, 434009.

[233] Matsunaga, Y., Li, C.-B., and Komatsuzaki, T. 2007. Anomalous diffusion in folding dynamics of minimalist protein landscape. *Physical Review Letters*, **99**, 238103.

[234] Matthiopoulos, J. 2003. The use of space by animals as a function of accessibility and preference. *Ecological Modelling*, **159**, 239–268.

[235] Mazzoni, A., Broccard, F. D., Garcia-Perez, E., *et al.* 2007. On the dynamics of the spontaneous activity in neuronal networks. *PloS ONE*, **5**, e439.

[236] Meats, A., and Edgerton, J. E. 2008. Short- and long-range dispersal of the Queensland fruit fly, *Bactrocera tryoni* and its relevance to invasive potential, sterile insect technique and surveillance trapping. *Australian Journal of Experimental Agriculture*, **48**, 1237–1245.

[237] Melloni, L., Schwiedrzik, C. M., Rodriguez, E., and Singer, W. 2009. (Micro)Saccades, corollary activity and cortical oscillations. *Trends in Cognitive Sciences*, **13**, 239–245.

[238] Meroz, Y., Eliazar, I., and Klafter, J. 2009. Facilitated diffusion in a crowded environment: from kinetics to stochastics. *Journal of Physics A*, **42**, 434012.

[239] Metzler, R. 2000. Generalized Chapman-Kolmogorov equation: A unifying approach to the description of anomalous transport in external fields. *Physical Review E*, **62**, 6233.

[240] Metzler, R., and Compte, A. 2000. Generalized diffusion-advection schemes and dispersive sedimentation: A fractional approach. *Journal of Physical Chemistry B*, **104**, 3858–3865.

[241] Metzler, R., and Klafter, J. 2000. The random walk's guide to anomalous diffusion: A fractional dynamics approach. *Physics Reports*, **339**, 1–77.

[242] Metzler, R., and Klafter, J. 2001. Lévy meets Boltzmann: Strange initial conditions for Brownian and fractional Fokker-Planck equations. *Physica A*, **302**, 290–296.

[243] Metzler, R., and Klafter, J. 2004. The restaurant at the end of the random walk: Recent developments in the description of anomalous transport by fractional dynamics. *Journal of Physics A*, **37**, R161–R208.

[244] Metzler, R., and Nonnenmacher, T. F. 2002. Space- and time-fractional diffusion and wave equations, fractional Fokker-Planck equations, and physical motivation. *Chemical Physics*, **284**, 67–90.

[245] Metzler, R., and Sokolov, I. M. 2002. Superdiffusive Klein-Kramers equation: Normal and anomalous time evolution and Lévy walk moments. *Europhysics Letters*, **58**, 482–488.

[246] Metzler, R., Ambjornsson, T., Hanke, A., Zhang, Y., and Levene, S. 2007. Single DNA conformations and biological function. *Journal of Computational and Theoretical Nanoscience*, **4**, 1–49.

[247] Metzler, R., Koren, T., van den Broek, B., Wuite, G. J. L., and Lomholt, M. A. 2009. And did he search for you, and could not find you? *Journal of Physics A*, **42**, 434005.

[248] Meysman, F. J. R., Malyuga, V. S., Boudreau, B. P., and Middelburg, J. J. 2008. A generalized stochastic approach to particle dispersal in soils and sediments. *Geochimica et Cosmochimica Acta*, **72**, 3460–3478.

[249] Mirny, L., Slutsky, M., Wunderlich, Z., *et al.* 2009. How a protein searches for its site on DNA: The mechanism of facilitated diffusion. *Journal of Physics A*, **42**, 434013.

[250] Montroll, E. W., and Weiss, G. 1965. Random walks on lattices. II. *Journal of Mathematical Physics*, **6**, 167–181.

[251] Moore, N. T., and Grosberg, A. Y. 2006. The abundance of unknots in various models of polymer loops. *Journal of Physics A*, **39**, 9081–9092.

[252] Morales, J. M., Haydon, D. T., Frair, J., Holsiner, K. E., and Fryxell, J. M. 2004. Extracting more out of relocation data: Building movement models as mixtures of random walks. *Ecology*, **85**, 2436–2445.

[253] Moreau, M., Bénichou, O., Loverdo, C., and Voituriez, R. 2007. Intermittent search processes in disordered medium. *Europhysics Letters*, **77**, 20006.

[254] Moreau, M., Bénichou, O., Loverdo, C., and Voituriez, R. 2009. Dynamical and spatial disorder in an intermittent search process. *Journal of Physics A*, **42**, 434007.

[255] Morse, P. M., and Kimball, G. E. 1951. *Methods of Operations Research*. New York: John Wiley.

[256] Morse, P. M., and Kimball, G. E. 1956. *How to Hunt a Submarine*, vol. 4, *The world of Mathematics*. New York, Simon and Schestor.

[257] Mueller, T., and Fagan, W. F. 2008. Search and navigation in dynamic environments – from individual behaviors to population distributions. *Oikos*, **117**, 654–664.

[258] Nara, S., Davis, P., and Totsuji, H. 1993. Memory search using complex dynamics in a recurrent neural network model. *Neural Networks*, **6**, 963.

[259] Nec, Y., Nepomnyashchy, A. A., and Golovin, A. A. 2008. Oscillatory instability in super-diffusive reaction-diffusion systems: Fractional amplitude and phase diffusion equations. *Europhysics Letters*, **82**, 58003.

[260] Nepomnyashchikh, V. A. 2003. The conflict between optimization and regularity in building behaviour of the caddisfly, *Chaetopteryx villosa Fabr. (Limnephilidae: Prichoptera)*, larvae. *Zhurnal Obshchei Biologii*, **64**, 45–54.

[261] Nepomnyashchikh, V. A., and Podgornyj, K. A. 2003. Emergence of adaptive searching rules from the dynamics of a simple nonlinear system. *Adaptive Behavior*, **11**, 245–265.

[262] Newlands, N. K., Lutcavage, M. E., and Pitcher, T. J. 2004. Analysis of foraging movements of Atlantic bluefin tuna (*Thunnus thynnus*): Individuals switch between two modes of search behaviour. *Population Ecology*, **46**, 39–53.

[263] Newman, J. R. 1956. *The World of Mathematics*. New York: Simon and Schuster.

[264] Nowak, M. A. 2006. *Evolutionary Dynamics: Exploring the Equations of Life*. Cambridge, MA: Harvard University Press.

[265] Okubo, A., and Levin, S. A. (eds.). 2001. *Diffusion and Ecological Problems: Modern Perspectives*. New York: Springer.

[266] Onsager, L. 1944. Crystal statistics. I. A two-dimensional model with an order-disorder transition. *Physical Review*, **65**, 117–149.

[267] Orians, G. H., and Pearson, N. E. 1979. On the theory of central place foraging. In *Analysis of Ecological Systems*, ed. D. J. Horn *et al.*, pp. 155–177. Columbus: Ohio State University Press.

[268] Oshanin, G., Wio, H. S., Lindenberg, K., and Burlatsky, S. F. 2007. Intermittent random walks for an optimal search strategy: One-dimensional case. *Journal of Physics – Condensed Matter*, **19**, 065142.

[269] Oshanin, G., Lindenberg, K., Wio, H. S., and Burlatsky, S. 2009. Efficient search by optimized intermittent random walks. *Journal of Physics A*, **42**, 434008.

[270] Pagurek, B., Dawes, N., and Kaye, R. 1992. A multiple paradigm diagnostic system for wide area communication networks. *Lecture Notes in Computer Science*, **604**, 256–265.

[271] Parashar, R., and Cushman, J. H. 2008. Scaling the fractional advective-dispersive equation for numerical evaluation of microbial dynamics in confined geometries with sticky boundaries. *Journal of Computational Physics*, **227**, 6598–6611.

[272] Pasternak, Z., Blasius, B., and Abelson, A. 2004. Host location by larvae of a parasitic barnacle: Larval chemotaxis and plume tracking in flow. *Journal of Plankton Research*, **26**, 487–493.

[273] Pasternak, Z., Blasius, B., Abelson, A., and Achituv, Y. 2006. Host-finding behaviour and navigation capabilities of symbiotic zooxanthellae. *Coral Reefs*, **25**, 201–207.

[274] Pasternak, Z., Bartumeus, F., and Grasso, F. W. 2009. Lévy-taxis: A novel search strategy for finding odor plumes in turbulent flow-dominated environments. *Journal of Physics A*, **42**, 434010.

[275] Patterson, T. A., Thomas, L., Wilcox, C., Ovaskainen, O., and Matthiopoulos, J. 2008. State-space models of individual animal movement. *Trends in Ecology and Evolution*, **23**, 87–94.

[276] Pauly, D., Christensen, V., Dalsgaard, J., Froese, R., and Torres, F., Jr. 1998. Fishing down marine food webs. *Science*, **279**, 860–863.

[277] Penrose, R. 1994. *Shadows of the Mind: A Search for the Missing Science of Consciousness*. Oxford: Oxford University Press.

[278] Pepin, D., Adrados, C., Mann, C., and Janeau, G. 2004. Assessing real daily distance traveled by ungulates using differential GPS locations. *Journal of Mammalogy*, **85**, 774–780.

[279] Pereira, E., Martinho, J. M. G., and Berberan-Santos, M. N. 2004. Photon trajectories in incoherent atomic radiation trapping as Lévy flights. *Physical Review Letters*, **93**, 120201.

[280] Peruani, F., and Morelli, L. G. 2007. Self-propelled particles with fluctuating speed and direction of motion in two dimensions. *Physical Review Letters*, **99**, 010602.

[281] Peters, R. 2005. Translocation through the nuclear pore complex: Selectivity and speed by reduction-of-dimensionality. *Traffic*, **6**, 421–427.

[282] Peterson, I. 1997. *The Jungles of Randomness: A Mathematical Safari*. New York: John Wiley.

[283] Petrovskii, S., Morozov, A., and Li, B.-L. 2008. On a possible origin of the fat-tailed dispersal in population dynamics. *Ecological Complexity*, **5**, 146–150.

[284] Phillips, R. A., Croxall, J. P., Silk, J. R. D., and Briggs, D. R. 2007. Foraging ecology of albatrosses and petrels from South Georgia: Two decades of insights from tracking technologies. *Aquatic Conservation – Marine and Freshwater Ecosystems*, **17**, S6–S21.

[285] Pinaud, D., and Weimerskirch, H. 2007. At-sea distribution and scale-dependent foraging behaviour of petrels and albatrosses: A comparative study. *Journal of Animal Ecology*, **76**, 9–19.

[286] Pirolli, P., and Card, S. 1995. Information foraging in information access environments. *Proceedings of the 1995 Conference on Human Factors in Computing Systems*, ed. G. C. vonder Veer and C. Gale, p. 381. New York: ACM Press.

[287] Pirolli, P. L. T. 2007. *Information Foraging Theory: Adaptive Interaction with Information*. New York: Oxford University Press.

[288] Plank, M. J., and James, A. 2008. Optimal foraging: Lévy pattern or process? *Journal of the Royal Society Interface*, **5**, 1077–1086.

[289] Plotnick, R. E. 2007. Chemoreception, odor landscapes, and foraging in ancient marine landscapes. *Palaeontologia Electronica*, **10**, 1A.

[290] Porto, M., and Roman, H. E. 2002. Autoregressive processes with exponentially decaying probability distribution functions: Applications to daily variations of a stock market index. *Physical Review E*, **65**, 046149.

[291] Radons, G., Klages, R., and Sokolov, I. M. (eds.) 2008. *Anomalous Transport*. Berlin: Wiley-VCH.

[292] Ramos-Fernández, G., Mateos, J. L., Miramontes, O., *et al.* 2004. Lévy walk patterns in the foraging movements of spider monkeys (*Ateles geoffroyi*). *Behavioral Ecology and Sociobiology*, **55**, 223–230.

[293] Randon-Furling, J., Majumdar, S. N., and Comtet, A. 2009. Convex hull of N planar Brownian motions: Exact results and an application to ecology. *Physical Review Letters*, **103**, 140602.

[294] Raposo, E. P., Buldyrev, S. V., da Luz, M. G. E., *et al.* 2003. Dynamical robustness of Lévy search strategies. *Physical Review Letters*, **91**, 240601.

[295] Raposo, E. P., Buldyrev, S. V., da Luz, M. G. E., Viswanathan, G. M., and Stanley, H. E. 2009. Lévy flights and random searches. *Journal of Physics A*, **42**, 434003.

[296] Reynolds, A. 2008. How many animals really do the Lévy walk? Comment. *Ecology*, **89**, 2347–2351.

[297] Reynolds, A. M. 2006. On the intermittent behaviour of foraging animals. *Europhysics Letters*, **75**, 517–520.

[298] Reynolds, A. M. 2006. Optimal scale-free searching strategies for the location of moving targets: New insights on visually cued mate location behaviour in insects. *Physics Letters A*, **360**, 224–227.

[299] Reynolds, A. M. 2007. *Preprint*.

[300] Reynolds, A. M. 2007. Avoidance of conspecific odour trails results in scale-free movement patterns and the execution of an optimal searching strategy. *Europhysics Letters*, **79**, 30006.

[301] Reynolds, A. M. 2008. Deterministic walks with inverse-square power-law scaling are an emergent property of predators that use chemotaxis to locate randomly distributed prey. *Physical Review E*, **78**, 011906.

[302] Reynolds, A. M. 2008. Optimal random Lévy-loop searching: New insights into the searching behaviours of central-place foragers. *Europhysics Letters*, **82**, 20001.

[303] Reynolds, A. M. 2009. Scale-free animal movement patterns: Lévy walks outperform fractional Brownian motions and fractional Lévy motions in random search scenarios. *Journal of Physics A*, **42**, 434006.

[304] Reynolds, A. M., and Frye, M. A. 2007. Free-flight odor tracking in *Drosophila* is consistent with an optimal intermittent scale-free search. *PLoS ONE*, **2**(4), e354.

[305] Reynolds, A. M., Reynolds, D. R., Smith, A. D., Svensson, G. P., and Lofstedt, C. 2007. Appetitive flight patterns of male *Agrotis segetum* moths over landscape scales. *Journal of Theoretical Biology*, **245**, 141–149.

[306] Reynolds, A. M., Smith, A. D., Menzel, R., *et al.* 2007. Displaced honey bees perform optimal scale-free search flights. *Ecology*, **88**, 1955–1961.

[307] Reynolds, A. M., Smith, D., Reynolds, D. R., Carreck, N. L., and Osborne, J. L. 2007. Honeybees perform optimal scale-free searching flights when attempting to locate a food source. *Journal of Experimental Biology*, **210**, 3763–3770.

[308] Reynolds, A. M. 2005. Scale-free movement patterns arising from olfactory-driven foraging. *Physical Review E*, **72**, 041928.

[309] Reynolds, A. M. 2006. Cooperative random Lévy flight searches and the flight patterns of honeybees. *Physics Letters A*, **354**, 384–388.

[310] Rhee, I., Shin, M., Hong, S., Lee, K., and Chong, S. 2008. On the Lévy-walk nature of human mobility. In *IEEE INFOCOM 2008 Proceedings*. Phoenix, Arizona: Curran Associates.

[311] Rhodes, C. J., and Anderson, R. M. 1997. Epidemic thresholds and vaccination in a lattice model of disease spread. *Theoretical Population Biology*, **52**, 101–118.

[312] Rhodes, T., and Turvey, M. T. 2007. Human memory retrieval as Lévy foraging. *Physica A*, **385**, 255–260.

[313] Riggs, T., Walts, A., Perry, N., *et al.* 2008. A comparison of random vs. chemotaxis-driven contacts of T cells with dendritic cells during repertoire scanning. *Journal of Theoretical Biology*, **250**, 732–751.

[314] Risau-Gusman, S., Martinez, A. S., and Kinouchi, O. 2003. Escaping from cycles through a glass transition. *Physical Review E*, **68**, 016104.

[315] Ritchie, M. E. 1998. Scale-dependent foraging and patch choice in fractal environments. *Evolutionary Ecology*, **12**, 309–330.

[316] Roman, H. E., and Porto, M. 2001. Self-generated power-law tails in probability distributions. *Physical Review E*, **6303**, 036128.

[317] Romero, P. D., and Candela, V. F. 2008. Blind deconvolution models regularized by fractional powers of the Laplacian. *Journal of Mathematical Imaging and Vision*, **32**, 181–191.

[318] Roshier, D. A., Doerr, V. A. J., and Doerr, E. D. 2008. Animal movement in dynamic landscapes: Interaction between behavioural strategies and resource distributions. *Oecologia*, **156**, 465–477.

[319] Royer, F., Fromentin, J. M., and Gaspar, P. 2005. A state-space model to derive bluefin tuna movement and habitat from archival tags. *Oikos*, **109**, 473–484.

[320] Santos, M. C., Raposo, E. P., Viswanathan, G. M., and da Luz, M. G. E. 2004. Optimal random searches of revisitable targets: Crossover from superdiffusive to ballistic random walks. *Europhysics Letters*, **67**, 734–740.

[321] Santos, M. C., Viswanathan, G. M., Raposo, E. P., and da Luz, M. G. E. 2005. Optimization of random searches on regular lattices. *Physical Review E*, **72**, 046143.

[322] Santos, M. C., Boyer, D., Miramontes, O., *et al.* 2007. Origin of power-law distributions in deterministic walks: The influence of landscape geometry. *Physical Review E*, **75**, 061114.

[323] Santos, M. C., Viswanathan, G. M., Raposo, E. P., and da Luz, M. G. E. 2008. Optimization of random searches on defective lattice networks. *Physical Review E*, **77**, 041101.

[324] Santos, M. C., Raposo, E. P., Viswanathan, G. M., and da Luz, M. G. E. 2009. Can collective searches profit from Levy walk strategies? *Journal of Physics A*, **42**, 434017.

[325] Saxton, M. 2008. A biological interpretation of transient anomalous subdiffusion. II. Reaction kinetics. *Biophysical Journal*, **94**, 760–771.

[326] Scher, H., and Lax, M. 1973. Stochastic transport in a disordered solid. I. Theory. *Physical Review B*, **7**, 4491–4502.

[327] Scher, H., and Montroll, E. W. 1975. Anomalous transit-time dispersion in amorphous solids. *Physical Review B*, **12**, 2455–2477.

[328] Schoener, T. W. 1971. Theory of feeding strategies. *Annual Review of Ecology and Systematics*, **2**, 369–404.

[329] Schooley, R. L., and Wiens, J. A. 2003. Finding habitat patches and directional connectivity. *Oikos*, **102**, 559–570.

[330] Schulz, M. 2002. Lévy flights in a quenched jump length field: A real space renormalization group approach. *Physics Letters A*, **298**, 105–108.

[331] Schulz, M., and Reineker, P. 2002. Lévy flights in a quenched jump length field: A 1-loop renormalization group approach. *Chemical Physics*, **284**, 331–340.

[332] Schuster, F. L., and Levandowsky, M. 1996. Chemosensory responses of *Acanthamoeba castellanii*: Visual analysis of random movement and responses to chemical signals. *Journal of Eukaryotic Microbiology*, **43**, 150–158.

[333] Scott, A. (ed.). 2005. *Encyclopedia of Nonlinear Science*. New York: Taylor and Francis.

[334] Segev, R., Benveniste, M., Hulata, E., *et al.* 2002. Long term behaviour of lithographically prepared in vitro neuronal networks. *Physical Review Letters*, **88**, 118102.

[335] Seuront, L. 2010. *Fractals and Multifractals in Ecology and Aquatic Science*. Boca Raton, FL: CRC Press.

[336] Seuront, L., Hwang, J. S., Tseng, L. C., *et al.* 2004. Individual variability in the swimming behavior of the sub-tropical copepod *Oncaea venusta* (Copepoda: Poecilostomatoida). *Marine Ecology – Progress Series*, **283**, 199–217.

[337] Seuront, L., Duponchel, A.-C., and Chapperon, C. 2007. Heavy-tailed distributions in the intermittent motion behaviour of the intertidal gastropod *Littorina littorea*. *Physica A*, **385**, 573–582.

[338] Shaw, M. W., Harwood, T. D., Wilkinson, M. J., and Elliott, L. 2006. Assembling spatially explicit landscape models of pollen and spore dispersal by wind for risk assessment. *Proceedings of the Royal Society B*, **273**, 1705–1713.

[339] Shlesinger, M. F. 2006. Mathematical physics – Search research. *Nature*, **443**, 281–282.

[340] Shlesinger, M. F. 2009. Random searching. *Journal of Physics A*, **42**, 434001.

[341] Shlesinger, M. F., and Klafter, J. 1986. Lévy walk versus Lévy flights. In *On Growth and Form*, ed. H. E. Stanley and N. Ostrowsky, pp. 279–283. Dordrecht, Netherlands: Martinus Nijhoff.

[342] Shlesinger, M. F., Zaslavsky, G. M., and Frisch, U. (eds.). 1995. *Lévy Flights and Related Topics in Physics*. Berlin: Springer.

[343] Shlesinger, M. F. 1974. Asymptotic solutions of continuous-time random walks. *Journal of Statistical Physics*, **10**, 421–434.

[344] Siegal, M. L., Promislow, D. E. L., and Bergman, A. 2007. Functional and evolutionary inference in gene networks: Does topology matter? *Genetica*, **129**, 83–103.

[345] Sims, D. W., Witt, M. J., Richardson, A. J., Southall, E. J., and Metcalfe, J. D. 2006. Encounter success of free-ranging marine predator movements across a dynamic prey landscape. *Proceedings of the Royal Society B*, **273**, 1195–1201.

[346] Sims, D. W., Righton, D., and Pitchford, J. W. 2007. Minimizing errors in identifying Lévy flight behaviour of organisms. *Journal of Animal Ecology*, **76**, 222–229.

[347] Sims, D. W., Southall, E. J., Humphries, N. E., *et al.* 2008. Scaling laws of marine predator search behaviour. *Nature*, **451**, 1098–1102.

[348] Sisterson, M. S., and Averill, A. L. 2002. Costs and benefits of food foraging for a braconid parasitoid. *Journal of Insect Behavior*, **15**, 571–588.

[349] Skellam, J. G. 1951. Random dispersal in theoretical populations. *Biometrika*, **38**, 196–218.

[350] Skoglund, H., and Barlaup, B. T. 2006. Feeding pattern and diet of first feeding brown trout fry under natural conditions. *Journal of Fish Biology*, **68**, 507–521.

[351] Smirnov, V. I. 1964. *A Course of Higher Mathematics, vol. 4, Integral Equations*. Oxford: Pergamon.

[352] Sogard, S. M., and Olla, B. L. 1996. Food deprivation affects vertical distribution and activity of a marine fish in a thermal gradient: Potential energy-conserving mechanisms. *Marine Ecology Progress Series*, **133**, 43–55.

[353] Sokolov, I. M. 2000. Lévy flights from a continuous-time process. *Physical Review E*, **63**, 011104.

[354] Sokolov, I. M., Chechkin, A., and Klafter, J. 2004. Fractional diffusion equation for a power-law-truncated Lévy process. *Physica A*, **336**, 245–251.

[355] Sole, R. V., Bartumeus, F., and Gamarra, J. G. P. 2005. Gap percolation in rainforests. *Oikos*, **110**, 177–185.

[356] Sparrevohn, C. R., Nielsen, A., and Stottrup, J. G. 2002. Diffusion of fish from a single release point. *Canadian Journal of Fisheries and Aquatic Sciences*, **59**, 844–853.

[357] Stanley, H. E. 1971. *Introduction to Phase Transitions and Critical Phenomena*. Oxford: Oxford University Press.

[358] Stanley, H. E., and Buldyrev, S. V. 2001. Statistical physics – The salesman and the tourist. *Nature*, **413**, 373–374.

[359] Stanley, H. E., and Ostrowsky, N. (eds.). 1986. *On Growth and Form*. Dordrecht, Netherlands: Martinus Nijhoff.

[360] Stanley, H. E., and Ostrowsky, N. (eds.). 1990. *Correlations and Connectivity: Geometric Aspects of Physics, Chemistry and Biology*. Dordrecht, Netherlands: Kluwer.

[361] Stanley, H. E., Amaral, L. A. N., Buldyrev, S. V., *et al.* 1996. Scaling and universality in animate and inanimate systems. *Physica A*, **231**, 20.

[362] Stanley, H. E., Amaral, L. A. N., Andrade, J. S., *et al.* 1998. Scale-invariant correlations in the biological and social sciences. *Philosophical Magazine B*, **77**, 1373–1388.

[363] Stanley, H. E., Amaral, L. A. N., Canning, D., *et al.* 1999. Econophysics: Can physicists contribute to the science of economics? *Physica A*, **269**, 156–169.

[364] Stephens, D. W., and Krebs, J. R. 1986. *Foraging Theory*. Princeton, NJ: Princeton University Press.

[365] Stephens, D. W., Brown, J. S., and Ydenberg, R. C. (eds.). 2007. *Foraging: Behavior and Ecology*. Chicago: University of Chicago Press.

[366] Sutherland, B. 2004. *Beautiful Models: 70 Years of Exactly Solved Quantum Many-Body Problems*. Singapore: World Scientific.

[367] Sutherland, W. J. 1996. *From Individual Behaviour to Population Ecology*. Oxford: Oxford University Press.

[368] Takahashi, H., Horibe, N., Shimada, M., and Ikegami, T. 2008. Analyzing the house fly's exploratory behavior with autoregression methods. *Journal of the Physical Society of Japan*, **77**, 084802.

[369] Tamura, K., Yusuf, Y., Hidaka, Y., and Kai, S. C. 2001. Nonlinear transport and anomalous Brownian motion in soft-mode turbulence. *Journal of the Physical Society of Japan*, **70**, 2805–2808.

[370] Thouless, C. R. 1995. Long distance movements of elephants in northern Kenya. *African Journal of Ecology*, **33**, 321–334.

[371] Thouless, C. R. 1996. Home ranges and social organization of female elephants in northern Kenya. *African Journal of Ecology*, **34**, 284–297.

[372] Tikhonov, D. A., Enderlein, J., Malchow, H., and Medvinsky, A. B. 2001. Chaos and fractals in fish school motion. *Chaos, Solitons and Fractals*, **12**, 277–288.

[373] Travis, J. 2008. How a shark finds its next meal. *ScienceNOW Daily News*. http://sciencenow.sciencemag.org/cgi/content/full/2008/227/1.

[374] Travis, J. M. J., and Palmer, S. C. F. 2005. Spatial processes can determine the relationship between prey encounter rate and prey density. *Biology Letters*, **1**, 136–138.

[375] Tsonis, A. A., Roebber, P. J., and Elsner, J. B. 1998. A characteristic time scale in the global temperature record. *Geophysical Research Letters*, **25**, 2821–2823.

[376] Tsonis, A. A., Hunt, A. G., and Elsner, J. B. 2003. On the relation between ENSO and global climate change. *Meteorology and Atmospheric Physics*, **84**, 229–242.

[377] Turchin, P. 1991. Translating foraging movements in heterogeneous environments into the spatial distribution of foragers. *Ecology*, **72**, 1253–1266.

[378] Turchin, P. 1998. *Quantitative Analysis of Movements: Measuring and Modeling Population Redistribution in Animals and Plants*. Sunderland, MA: Sinauer Associates.

[379] Uttieri, M., Cianelli, D., Strickler, J. R., and Zamblanchi, E. 2007. On the relationship between fractal dimension and encounters in three-dimensional trajectories. *Journal of Theoretical Biology*, **247**, 480–491.

[380] van Dartel, M., Postma, E., van den Herik, J., and de Croon, G. 2004. Macroscopic analysis of robot foraging behaviour. *Connection Science*, **16**, 169–181.

[381] van der Veer, G. C., and Gale, C. (eds.). 1995. *Proceedings of the 1995 Conference on Human Factors in Computing Systems*. New York: ACM Press.

[382] van Gils, J. A., Piersma, T., Dekinga, A., Spaans, B., and Kraan, C. 2006. Shellfish dredging pushes a flexible avian top predator out of a marine protected area. *PLoS Biology*, **4**, e376.

[383] Van Houtan, K. S., Pimm, S. L., Halley, J. M., Bierregaard, R. O., Jr., and Lovejoy, T. E. 2007. Dispersal of Amazonian birds in continuous and fragmented forest. *Ecology Letters*, **10**, 219–229.

[384] Vazquez, A., Oliveira, J. G., Dezso, Z., *et al.* 2006. Modeling bursts and heavy tails in human dynamics. *Physical Review E*, **73**, 036127.

[385] Vicsek, T. 1992. *Fractal Growth Phenomena*. Singapore: World Scientific.

[386] Vicsek, T., Czirókl, A., Ben-Jacob, E., Cohen, I., and Shochet, O. 1995. Novel type of phase transition in a system of self-driven particles. *Physical Review Letters*, **75**, 1226–1229.

[387] Visser, A. W. 2007. Motility of zooplankton: Fitness, foraging and predation. *Journal of Plankton Research*, **29**, 447–461.

[388] Visser, A. W., and Kiorboe, T. 2006. Plankton motility patterns and encounter rates. *Oecologia*, **148**, 538–546.

[389] Viswanathan, G. M., and Viswanathan, T. M. 2008. Spontaneous symmetry breaking and finite-time singularities in *d*-dimensional incompressible flows with fractional dissipation. *Europhysics Letters*, **84**, 50006.

[390] Viswanathan, G. M., Afanasyev, V., Buldyrev, S. V., *et al.* 1996. Lévy flight search patterns of wandering albatrosses. *Nature*, **381**, 413–415.

[391] Viswanathan, G. M., Peng, C.-K., Stanley, H. E., and Goldberger, A. L. 1997. Deviations from uniform power law scaling in nonstationary time series. *Physical Review E*, **55**, 845–849.

[392] Viswanathan, G. M., Buldyrev, S. V., Havlin, S., and Stanley, H. E. 1997. Quantification of DNA patchiness using long-range correlation measures. *Biophysical Journal*, **72**, 866–875.

[393] Viswanathan, G. M., Buldyrev, S. V., Havlin, S., *et al.* 1999. Optimizing the success of random searches. *Nature*, **401**, 911–914.

[394] Viswanathan, G. M., Afanasyev, V., Buldyrev, S. V., *et al.* 2001. Lévy flights search patterns of biological organisms. *Physica A*, **295**, 85–88.

[395] Viswanathan, G. M., Afanasyev, V., Buldyrev, S. V., *et al.* 2001. Statistical physics of random searches. *Brazilian Journal of Physics*, **31**, 102–108.

[396] Viswanathan, G. M., Raposo, E. P., Bartumeus, F., Catalan, J., and da Luz, M. G. E. 2005. Necessary criterion for distinguishing true superdiffusion from correlated random walk processes. *Physical Review E*, **72**, 011111.

[397] Viswanathan, G. M., Raposo, E. P., and da Luz, M. G. E. 2008. Lévy flights and superdiffusion in the context of biological encounters and random searches. *Physics of Life Reviews*, **5**, 133–150.

[398] Walther, H. 1999. Spectroscopy of single trapped ions and applications to frequency standards and cavity quantum electrodynamics. *Laser Physics*, **9**, 225–233.

[399] Watts, D. J., and Strogatz, S. H. 1998. Collective dynamics of "small-world" networks. *Nature*, **393**, 440–442.

[400] Waugh, S., Filippi, D., Fukuda, A., *et al.* 2005. Foraging of royal albatrosses, *Diomedea epomophora*, from the Otago peninsula and its relationships to fisheries. *Canadian Journal of Fisheries and Aquatic Sciences*, **62**, 1410–1421.

[401] Weimerskirch, H. 2007. Are seabirds foraging for unpredictable resources? *Deep-Sea Research, Part II*, **54**, 211–223.

[402] Weimerskirch, H., and Guionnet, T. 2002. Comparative activity pattern during foraging of four albatross species. *Ibis*, **144**, 40–50.

[403] Weimerskirch, H., Guionnet, T., Martin, J., Shaffer, S. A., and Costa, D. P. 2000. Fast and fuel efficient? Optimal use of wind by flying albatrosses. *Proceedings of the Royal Society B*, **267**, 1869–1874.

[404] Weimerskirch, H., Gault, A., and Cherel, Y. 2005. Prey distribution and patchiness: Factors in foraging success and efficiency of wandering albatrosses. *Ecology*, **86**, 2611–2622.

[405] Wells, K., Pfeiffer, M., Lakim, M. B., and Kalko, E. K. V. 2006. Movement trajectories and habitat partitioning of small mammals in logged and unlogged rain forests on Borneo. *Journal of Animal Ecology*, **75**, 1212–1223.

[406] Wells, K., Kalko, E. K. V., Lakim, M. B., and Pfeiffer, M. 2008. Movement and ranging patterns of a tropical rat (*Leopoldamys sabanus*) in logged and unlogged rain forests. *Journal of Mammalogy*, **89**, 712–720.

[407] West, B. J., Hamilton, P., and West, D. J. 1999. Fractal scaling of the teen birth phenomenon. *Fractals – Complex Geometry Patterns and Scaling in Nature and Society*, **7**, 113–122.

[408] Westcott, D. A., and Graham, D. L. 2000. Patterns of movement and seed dispersal of a tropical frugivore. *Oecologia*, **122**, 249–257.

[409] White, E. P., Enquist, B. J., and Green, J. L. 2008. On estimating the exponent of power-law frequency distributions. *Ecology*, **89**, 905–912.

[410] Wiersma, D. S. 2008. The physics and applications of random lasers. *Nature Physics*, **4**, 359–367.

[411] Wilson, R. P., Liebsch, N., Davies, I. M., *et al.* 2007. All at sea with animal tracks; methodological and analytical solutions for the resolution of movement. *Deep-Sea Research, Part II*, **54**, 193–210.

[412] Worrall, D. R., Kirkpatrick, I., and Williams, S. L. 2004. Bimolecular processes on silica gel surfaces: Energetic factors in determining electron-transfer rates. *Photochemical and Photobiological Sciences*, **3**, 63–70.

[413] Yuste, S. B., and Acedo, L. 1999. Territory covered by N random walkers. *Physical Review E*, **60**, R3459–R3462.

[414] Zanette, D. H. 1999. Statistical-thermodynamical foundations of anomalous diffusion. *Brazilian Journal of Physics*, **29**, 108–124.

[415] Zanette, D. H., and Montemurro, M. A. 2003. Thermal measurements of stationary nonequilibrium systems: A test for generalized thermostatistics. *Physics Letters A*, **316**, 184–189.

[416] Zaslavsky, G. M. 2002. Chaos, fractional kinetics, and anomalous transport. *Physics Reports*, **371**, 461–580.

[417] Zhang, X., Johnson, S. N., Crawford, J. W., Gregory, P. J., and Young, I. M. 2007. A general random walk model for the leptokurtic distribution of organism movement: Theory and application. *Ecological Modelling*, **200**, 79–88.

[418] Zurek, W. H. 2003. Decoherence, einselection, and the quantum origins of the classical. *Reviews of Modern Physics*, **75**, 715–775.

Index